Workplace Environmental Factors

This concise volume examines how environmental conditions, specifically lighting, noise, and vibration, affect human performance, health, and comfort. As workplace demands evolve, so must our approach to design. This book covers the importance of aligning visual and auditory elements of work environments with the needs and limitations of the human body, whilst also enhancing productivity and minimizing risk.

This book details the tools and insights needed to better support human well-being in complex environments. Readers will discover thorough coverage of lighting principles, from basic definitions and equipment types to contemporary developments in visual display technologies and their ergonomic implications. The book also presents an in-depth treatment of noise in the workplace, covering measurement, physiological effects, and protective strategies, as well as an analysis of human exposure to vibration. Together, these chapters provide readers with a practical understanding of how to evaluate and improve the sensory and physical quality of working environments.

Workplace Environmental Factors: Lighting, Noise, and Vibration is essential for professionals in ergonomics and human factors engineering, as well as for industrial designers, occupational health specialists, and policymakers involved in shaping safe and effective workspaces.

Body of Knowledge in Human Factors and Ergonomics

Series Editor: Waldemar Karwowski, University of Central Florida, USA, and Beata Mrugalska, Poznan University of Technology, Poznan, Poland

This Focus book series aims to provide the perspective readers-at-large with basic knowledge of human factors and ergonomics that can be useful to students, teachers, academicians, researchers, and practitioners in domains of human activity, across all sectors of industry, business, education, home, leisure, entertainment, and sports. This new Focus series will cover the whole spectrum of topics related to a broadly defined HF/E field. All potential chapters will be selected from the content of the HF/E encyclopedia, built out and arranged into thematic short form Focus titles on the specific aspects of research, practice, and education within this field.

Should you be interested in including your own book in this series, please contact James Hobbs the Editor for Ergonomics/Human Factors and Occupational Health and Safety on james.hobbs@tandf.co.uk.

Anthropometry: Human Body Measurements and How to Use Them
Edited by Beata Mrugalska and Waldemar Karwowski

Human Factors in Challenging Environments
Edited by Waldemar Karwowski and Beata Mrugalska

Workplace Environmental Factors
Edited by Waldemar Karwowski and Beata Mrugalska

For more information about this series, please visit: https://www.routledge.com/Body-of-Knowledge-in-Human-Factors-and-Ergonomics/book-series/CRCFBK

Workplace Environmental Factors

Lighting, Noise, and Vibration

Edited by
Waldemar Karwowski and Beata Mrugalska

Routledge
Taylor & Francis Group
NEW YORK AND LONDON

First published 2026
by Routledge
4 Park Square, Milton Park, Abingdon, Oxon OX14 4RN

and by Routledge
605 Third Avenue, New York, NY 10158

Routledge is an imprint of the Taylor & Francis Group, an informa business

ISBN: 9781041141556 (hbk)
ISBN: 9781041147138 (pbk)
ISBN: 9781003675761 (ebk)

DOI: 10.1201/9781003675761

Typeset in Times
by Deanta Global Publishing Services, Chennai, India

Contents

About the Editors

Beata Mrugalska is an Associate Professor, Vice-Dean for Science at the Faculty of Management Engineering (2024–2028) and Head of the Division of Applied Ergonomics (2020–2024, 2024–2028), Institute of Safety and Quality Engineering, Faculty of Management Engineering, Poznan University of Technology, Poland. She holds an M.Sc. in Management and Marketing, and a Ph.D. in Machine Construction and Operation, and was awarded a D.Sc. in Mechanical Engineering from Poznan University of Technology, Poland. Since 2018, she has been a board member of the Center for Registration of European Ergonomists (CREE). She is responsible for promoting the professional title of EuroErgonomist in the international scientific and industrial environment. She is also a board member of the Polish Academy of Arts and Sciences, Commission of Ergonomics, and a member of the Ergonomics Committee at the Polish Academy of Sciences, Branch Poznan. She has over 150 publications on human factors in modern organizational management concepts.

Waldemar Karwowski is a Pegasus Professor and Chairman of the Department of Industrial Engineering and Management Systems, and Executive Director of the Institute for Advanced Systems Engineering, University of Central Florida, United States. He holds an M.S. in Production Engineering and Management from the Technical University of Wroclaw, Poland, and a Ph.D. in Industrial Engineering from Texas Tech University, United States. He was awarded a D.Sc. degree in Management Science by the State Institute for Organization and Management in Industry, Poland. He is Past President of the International Ergonomics Association (2000–2003) and the Human Factors and Ergonomics Society (2007). He served on the Committee of Human Factors/Human Systems Integration, National Research Council, and the National Academies, United States (2007–2011). He has over 500 publications focused on human performance, safety, cognitive systems engineering, human-centered design, neuro-fuzzy systems, nonlinear dynamics, neurotechnology, and neuroergonomics. He serves as the Co-Editor-in-Chief of the journal *Theoretical Issues in Ergonomics Science* (Taylor & Francis).

List of Contributors

D. Augustynska
Katedra Mechaniki i Wibroakustyki, Wydzial Inżynierii Mechanicznej i Robotyki, Akademia Górniczo-Hutnicza
Krakow, Poland

P. R. Boyce
Lighting Research Center, Rensselaer Polytechnic Institute, New York
United States

P. L. Chong
School of Engineering and Computing, MILA University, Putra Nilai
Malaysia

Z. Engel
Katedra Mechaniki i Wibroakustyki, Wydzial Inżynierii Mechanicznej i Robotyki, Akademia Górniczo-Hutnicza
Krakow, Poland

J. M. Hesse
Ergonomics Division/University of Siegen, Siegen
Germany

H. Irle
Ergonomics Division/University of Siegen, Siegen
Germany

A. Kaczmarska
Katedra Mechaniki i Wibroakustyki, Wydzial Inżynierii Mechanicznej i Robotyki, Akademia Górniczo-Hutnicza
Krakow, Poland

F. Y. Kong
Faculty of Engineering and Technology
Malaysia

J. Koton
Katedra Mechaniki i Wibroakustyki, Wydzial Inżynierii Mechanicznej i Robotyki, Akademia Górniczo-Hutnicza
Krakow, Poland

R. Burgess-Limerick
School of Human Movement Studies, The University of Queensland
Australia

H. Luczak
Institute of Industrial Engineering and Ergonomics, RWTH Aachen University
Germany

P. Manokaran
School of Engineering and Computing, MILA University, Putra Nilai
ZTE (Malaysia) Corporation, Selangor
Malaysia

B. Mrugalska
Poznan University of Technology, Poznan
Poland

P. K. Ng
Faculty of Engineering and Technology, Multimedia University
Malaysia

O. Oehme
Institute of Industrial Engineering and Ergonomics, RWTH Aachen University
Germany

D. L. Post
Air Force Research Laboratory, Wright-Patterson Air Force Base
United States

M. Rötting
Liberty Mutual Research Center for Safety and Health
United States

L. Schmidt
Institute of Industrial Engineering and Ergonomics, RWTH Aachen University
Germany

P. J. School

H. Strasser
Ergonomics Division/University of Siegen, Siegen
Germany

H. L. Task
Air Force Research Laboratory, Wright-Patterson Air Force Base
United States

E. Tytyk
Poznan University of Technology, Poznan
Poland

D. E. Wasserman
Human Vibration and Biomedical Engineering Consultant, Cincinnati
United States

A. Wolska
Safety Engineering Department, Central Institute for Labour Protection – National Research Institute,Warsaw
Poland

Preface

Today, human factors and ergonomics professionals worldwide contribute to the design and evaluation of tasks, jobs, products, environments, and systems in order to make them compatible with the needs, abilities, and limitations of people (IEA 2000). These professionals, who are often certified by the professional certification bodies (e.g. Board of Certification in Professional Ergonomics in the United Sates, or the Centre for Registration of European Ergonomists – CREE in the European Union), promote the human-centered approach to work systems design, testing, and evaluation, which considers the broad set of physical, cognitive, social, organizational, environmental, and other relevant human factors. Such actions should help in the socio-economic development of the world society at large. As noted by the National Academy of Engineering in the United States, in the near future, "engineering will expand toward tighter connections between technology and the human experience …, and ergonomic design of engineered products" (NAE 2004). For example, the International Ergonomics Association works to enhance the public understanding of the meaning of ergonomics, and facilitates making informed decisions about the promotional claims of "ergonomically designed" systems. One of its elements is environmental ergonomics, which refers to our physiological and behavioral reactions to the ambient environment, and the design of effective barriers that allow us to survive in inhospitable settings.

This book includes not only chapters on environmental conditions but also discusses ergonomics in the context of many physical industrial settings. Its subject areas include lighting, noise, and vibration.

REFERENCE

IEA, 2000, International Ergonomics Association: www.iea.cc. NAE, 2004, The Engineer of 2020: Visions of Engineering in the New Century. Washington, DC: National Academy of Engineering, National Academies Press.

1 Human Aspects of Lighting in Working Interiors

A. Wolska

1.1 INTRODUCTION

Lighting is one of the main environmental factors in the work area. Over 80% of information about the surroundings is obtained through the eyes. Good lighting is required for good visibility of the environment. Lighting should ensure a luminous environment that is human-friendly and appropriate for the visual tasks performed. Therefore, visual strain should be minimized and well-being ensured during visual work under a given lighting system. Good lighting practice is not just limited to compliance with the requirements of standards but also should take into consideration the lighting preferences of users and the characteristics of those users, such as age or gender (Wolska 2003). In general, the purpose of lighting is to ensure the following:

- Visual comfort
- Visual performance
- Visual safety

DOI: 10.1201/9781003675761-1

Safety should be ensured in all circumstances, but different emphasis is placed on visual performance and the appearance of the interior, depending on the nature of the interior and type of task or activity. For example, lighting for a tool room in a factory should place far more emphasis on lighting the task and on visual performance than on the pleasant appearance of the room, but in a restaurant or a hotel room, the priorities should be reversed. In all cases, energy consumption and cost-effectiveness of a lighting system should be considered.

1.2 LIGHTING TERMINOLOGY (PARAMETERS)

1.2.1 Light (Visible Radiation)

Light is radiant energy that is capable of exciting the human retina and creating a visual sensation (Rea 1993). Light is visible radiation, and both terms can be used interchangeably. The commonly used spectrum of visible radiation is the range between 380 and 780 nm on the electromagnetic spectrum, although the limits are not defined precisely and depend on the radiant flux that is incident on the retina and the spectral sensitivity of the observer. In general, the lower limit is between 360 and 400 nm, and the upper limit is between 760 and 830 nm (Figure 1.1).

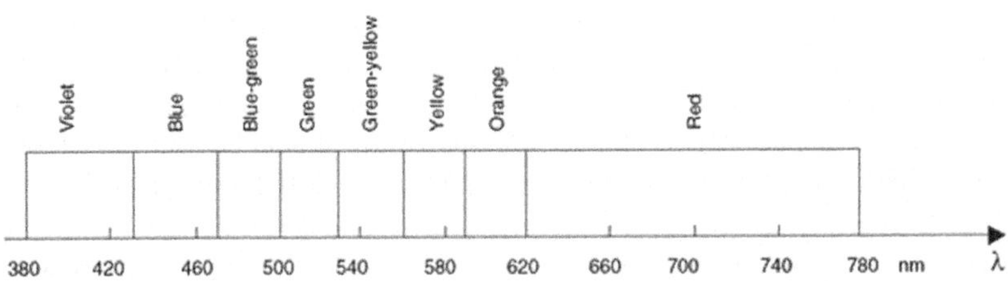

FIGURE 1.1 Subranges of visible radiation. The visible spectrum (from 380 nm to 780 nm) can be divided into eight subranges, which correspond to the relevant colors of light: violet, blue, blue-green, green, yellow-green, yellow, orange, and red.

1.2.2 Spectral Luminous Efficiency Curves

Physical quantities describing the radiation (such as radiant flux, irradiance, radiant intensity) are insufficient to define the efficiency of radiation to create a visual sensation.

The sensitivity of the human eye is different for individual wavelengths of visible radiation. This means that the eye is more sensitive to some wavelengths than to others. There are individual variations, and not everyone has the same sensitivity. However, the standardized sensitivity of the human eye for individual wavelengths of the whole range of visible radiation was needed for the purpose of lighting calculations and measurements. The International Commission on Illumination (CIE) adopted and agreed on standard responses, called the CIE Standard Photopic Observer and CIE Standard Scotopic Observer, and established luminous efficiency curves for photopic vision $V(\lambda)$ and scotopic vision $V(\lambda)$ (Figure 1.2).

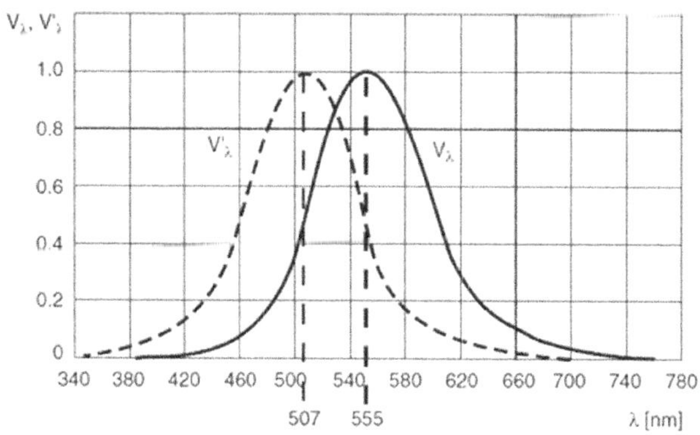

FIGURE 1.2 Luminous efficiency curves for photopic and scotopic vision. The spectral luminous efficiency function for photopic (cones) vision has maximum sensitivity at 555 nm (green-yellow) and for scotopic (rods) vision has maximum sensitivity at 507 nm (blue-green). These curves are the basis of the conversion from radiometric to photometric quantities.

1.2.3 Photometric Quantities

Photometric quantities are the physical quantities assessed on the basis of the visual sensation they create.

1. *Luminous flux* (unit: lm, lumen), a quantity derived from radiant flux (in watts, W) by evaluating the radiation according to the spectral sensitivity of the human eye (as defined by the CIE standard photometric observer) (CEN 2002b).
2. *Luminous intensity I* (unit: cd, candela), the angular density of the luminous flux in a given direction. This is luminous flux emitted in a very narrow (unit) solid angle containing the given direction. One candela is equal to one lumen per steradian (1 cd 1 lmsr –1).
3. *Illuminance E* (unit: lx, lux), the area density of a luminous flux that is incident at a point on a surface (luminous flux per unit area, lx = lmm–2).
4. *Luminance L* (unit: cdm–2 candelas per square meter), physical measure of brightness. It represents the luminous intensity emitted (or reflected) in a given direction per projected area of a luminous (or reflecting) surface.
5. *Contrast C*, the luminance difference between two objects, surfaces, or parts of the objects in the visual field, which are adjacent to each other (viewed simultaneously) or observed one after the other (successively). Contrast is defined in several ways:

    - *Luminance ratio*: ratio between the luminances of any two areas in the visual field (usually observed successively):

$$C = \frac{L_1}{L_2} \tag{1.1}$$

    - *Luminance contrast*: relationship between the luminance of an object and that of its immediate background (for surfaces viewed simultaneously):

$$C = \frac{L_1 - L_2}{L_1} \tag{1.2}$$

where

$L_1 > L_2$

- *Luminance factor LF*: ratio of the luminance of the surface viewed from a specific position and in a specified way to the luminance of a diffusely reflecting white surface from the same direction and lit in the same way (Boyce 2003).
- *Reflectance*: ρ , the physical property of a surface. Reflectance is the ratio of the luminous flux reflected from a given surface to the luminous flux incident on it. Reflectance values that approximately equal 1 correspond to perfectly white surfaces, and values that almost equal 0 correspond to perfectly black surfaces. Three main kinds of reflection can be distinguished depending on the type of texture of the surface: specular (mirror or polished surfaces), diffuse (matte surfaces), and mixed, which is combination of specular and diffuse reflection (semimatte surfaces, such as silk or polished scratches on a matte surface).

For matte surfaces (diffuse reflection), the following relationship can be used:

$$L = \frac{\rho * E}{\pi} \tag{1.3}$$

where L is luminance (cd/m2); ρ is reflectance of the surface considered; E is illuminance (lx); $\pi = 3.14$.

For a nondiffusely reflecting surface, reflectance in Equation 1.3 is replaced with the luminance factor.

1.3 VISUAL PERFORMANCE AND LIGHTING

There are many different types of work stands in industry, offices, and in other working interiors, which require specific visual conditions to perform visual tasks, and which are often very complex. The objective of lighting design is to minimize strain of the visual system and to achieve efficient visual performance. Prolonged visual work in poor lighting conditions induces visual fatigue, which can manifest itself in a variety of asthenopic symptoms, such as headache, redness, tender, itchy, burning, dry, and throbbing eyes, focusing problems, and blurred or double vision. Visual fatigue symptoms are temporary, and after a rest period, these symptoms should disappear. The optimum values of lighting parameters are determined with the use of a visual performance assessment, in terms of speed and accuracy of the performance of visual tasks. Visual performance depends both on the visual capability of the observer and visibility of the task. Lighting engineering can influence only one group of factors that are connected with the visibility of the task. The visibility of task depends on two groups of factors:

- *Task attributes*: size, distance, luminance, contrast, color, and surface properties.
- *Illumination*: illuminance level, illuminance uniformity, luminance distribution, spectral content of light sources, and glare.

Knowledge of the task attributes is essential so that the lighting designer can prepare an appropriate design of the lighting environment.

1.4 LIGHTING CHARACTERISTICS

Interior lighting conditions are described by the following lighting parameters, which determine whether the luminous environment is satisfactory:

- Illuminance level
- lluminance uniformity
- Luminance distribution
- Glare
- Spectral characteristic of light
- Flicker

1.4.1 Illuminance Level

The illuminance level that is required for the performance of visual tasks depends on two parameters:

- Apparent target size (function of the smallest size of the target and the viewing distance)
- Degree of visual task difficulty (depends on task contrast and reflectance)

The degree of visual task difficulty increases when the task contrast and reflectance decrease. Greater visual difficulty and a smaller apparent target size require a higher illuminance level. An example of a very difficult visual task is sewing black fabric (cloth) using black thread, because it represents a small size of target (stitch dimensions), low reflectance (close to 0), and poor contrast (close to 0, using Equation 1.2).

A relationship exists between illumination on the task and visual performance, expressed by speed and accuracy of test performance. An increase in the illumination produces an increase in performance, especially for visually difficult tasks. The greater the visual difficulty, the greater the effect of the illuminance; however, increasing the illumination does not make the level of visual performance of a difficult visual task equal to that of an easy visual task. Larger improvements in performance of visually difficult tasks can be achieved only by increasing the task size and contrast.

In practice, the required illuminance level for a given task refers to the average illuminance on the task area and the surrounding area or on the working plane. Average illuminance on a given surface is calculated as the arithmetic mean of at least three measured illuminances at evenly distributed points on that surface. Because of the trend towards the same lighting requirements in all countries, both the European standard (CEN 2002a) and the international standard (ISO 2002) have implemented the requirements of the International Commission on Illumination Standard (CIE 2001). All the above lighting standards have introduced a definition of maintained illuminance. This is the value of illuminance below which the average illuminance on the specified surface should not fall.

According to Weber's law and lighting practice, which refer to the smallest increment needed to give a significant difference in the subjective lighting effect, the recommended scale of illuminance is (CIE 2001): 20–30–50–75–100–150–200–300–500–750–1000–1500–2000–3000–5000 lx. The requirements of maintained illuminance for different areas, tasks, or activities presented in the standards are valid for normal visual conditions and take into account the degree of visual task difficulty, visual comfort, economy, and safety (see Table 1.1). In continuously occupied areas, 200 lx is recommended as the minimum maintained illuminance.

1.4.2 Illuminance Uniformity

Variation of illuminance should be considered in two areas: on and around the visual task area, and in the whole interior. Excessive changes of illuminance in the visual field can be distracting, and too much nonuniformity in circulation areas can cause poor visibility of any obstacles and ultimately accidents.

TABLE 1.1 Examples of Recommended Maintained Illuminance (Selected from CIE 2001)

MAINTAINED ILLUMINANCE, LX	*EXAMPLES OF INTERIORS OR ACTIVITIES*
50	Automatic processing, remotely operated processing installations
75	Indoor parking areas
100	Circulation areas, rest rooms, stores, and stockrooms
150	Stairs, escalators, processing installations with limited manual intervention
200	Machine shops, waiting rooms, canteens, archives, rough assembly
300	Continuously manned workplaces in processing installations, general machine work, galvanizing, washing, ironing, welding, rough and average machining, copying, exhibition halls, reception desk, medium assembly, classrooms
500	Reading, writing, conference rooms, data processing, precision machining, fine assembly, laboratories, rooms for medical attention, control rooms, lecture hall, kitchen
750	Precision assembly, sewing, technical drawing, grinding, scribing, polishing, inspection
1000	Quality control, final inspection, color inspection, fabric control, medical examination and treatment, lithography, typesetting, manufacture of jewelry, decorating, grinding, micromechanics
1500	Working with precious stones, watch-making, electronic workshops, color inspection in multicolored printing, invisible mending
2000	Steel and copper engraving
5000	Operating cavities and matching white teeth at dental surgeries, autopsy table

Uniformity of illuminance is calculated as the ratio of minimum measured illuminance to the average illuminance of a given working plan (task area, immediate surrounding area). It should be pointed out that the task area may be smaller than the entire area of the workplace, and this should be taken into account during the measurements. As a result, there is a distinction between the task area and the immediate surrounding area, which is a minimum of a 0.5 m band surrounding the task area within the field of vision. The recommended illuminance uniformity should not be less than (CIE 2001):

- 0.7 on the task area
- 0.5 on the immediate surrounding area

In general, the illuminance should change gradually. In work interiors with tasks of different illuminance requirements, it is not economically reasonable to illuminate the whole interior to a higher level of illuminance. In such cases, general lighting can provide a lower level of illuminance, and only tasks of greater visual difficulty should be additionally lit by localized or local lighting. However, in such cases, it is important for the average illuminance difference between two adjacent interiors (corridor and room) and between adjacent working planes in the same room to be no greater than 1:5.

1.4.3 Luminance Distribution

The luminance distribution in the field of view controls the adaptation level of the eyes, which affects task visibility. The luminance distribution in the field of view also affects visual comfort. For a given lighting level, differences in luminance will be caused by differences in surface reflectance. It can happen that the illuminance is appropriate for the visual task, but the luminance balance in the whole of the interior is unacceptable. The luminance distribution can be described either by luminance ratios of adjacent surfaces and surfaces viewed in sequence or by reflectances of major interior surfaces. Excessively high luminance ratios can cause fatigue and/or glare, but excessively low luminance ratios generally result in a dull and nonstimulating working environment.

It is recommended that the luminance ratio of task areas that are frequently viewed in sequence or adjacent surfaces should be lower than 10:1. In order to improve visual performance, the luminance of the task environment should be lower than task luminance, but no less than 1:3. There are also recommendations for reflectances of major interior surfaces (CIE 2001), which usually give the ranges of the preferred values for ceilings (0.6–0.9), walls (0.3–0.8), working planes (0.2–0.6), and the floor (0.1–0.5). The recommended values of working plane reflectances indicate that those surfaces should neither be very bright (white) nor very dark (black or dark brown), and this should be considered before furnishing the working interiors.

1.4.4 Glare

The lighting system should provide a luminous environment that is free of glare. This phenomenon exists when parts of the visual field are excessively bright in relation to the brightness of the general surroundings to which eyes are adapted. Glare is experienced when the excessive brightness of an object (light source) is in or close to the direction of the line of sight (direct glare) or when such light sources are reflected in glossy surfaces (reflected glare). From the psychophysiological point of view, there are two kinds of glare: disability glare and discom fort glare.

1.4.5 Disability Glare

Disability glare impairs the vision of details or objects (visibility and visual performance are reduced) without necessarily causing discomfort (CEN 2002b). It occurs when a large source of low luminance or a small source of high luminance is seen close to the line of sight to the visual task. Disability glare involves a veiling luminance produced in the eye on the retina by scattered light in ocular media inhomogeneities. This luminance is superimposed on the well-focused image on the retina and, as a result, reduces its contrast, resulting in reduced visual per- performance and visibility. Veiling luminance, or in other words disability glare, depends on illuminance from the glare source at the observer eye and the angle between the primary object and the glare sources. It is largely independent of the luminance of the source, with some effect from the time of exposure.

1.4.6 Discomfort Glare

Discomfort glare causes a feeling of discomfort, annoyance, and irritation without necessarily impairing the vision of details or objects. It increases with time and may contribute to fatigue. It has been found that discomfort glare depends on the luminances of the glare sources, the adaptation luminance level (background luminance), the number of glare sources, the solid angle subtended by the source at the observer's eye, and the angular displacement of the glare sources from the observer's line of sight. Discomfort glare must be controlled at the design stage, and, in fact, this is the control of the luminances of lamps in the direction of the observer's eyes. The degree of discomfort glare also depends on the kind of activity being performed. A more demanding visual task will result in a stronger feeling of discomfort and will need more restricted luminance control requirements. The limitation of discomfort glare is one of the major quality aspects of lighting because when discomfort glare is properly controlled, disability glare does not exist.

There are many different glare-limiting systems, which have as their main assumption the prediction of discomfort glare in the working environment, but most of them can only be used in a limited number of conditions. The discomfort glare evaluation system recommended for use by the CIE is the unified glare rating (UGR) tabular method, the practical application of which requires the use of computer software. This method is based on the following formula (CIE 2001):

$$UGR = 8\log(\frac{0.25}{L_b}\sum\frac{L^2 * \omega}{p^2}) \tag{1.4}$$

where L_b is background luminance; L, luminance of the bright parts of lights in the direction of the observer's eye; ω, solid angle of the bright parts of lights at the observer's eye; and p, Guth position index.

The *UGR* value calculated for the designed lighting installation should not exceed the values given in the standards by type of interior or activity. The *UGR* scale is: 13, 16, 19, 22, 25, 28, where each step in the scale represents one significant change in glare effect, and 13 represents the least perceptible discomfort glare. The maximum permissible design *UGR* values for lighting installations are provided according to the type of interior, task, or activity. In practice, 16 is the lowest *UGR* value required in the standards, and it should be assured for tasks with difficult and special requirements, such as the manufacture of synthetic precious stones, color inspection, or electronic workshops. The lowest glare limitation (*UGR* 28) is recommended for simple tasks, circulation areas, and interiors not used continuously for work purposes.

In practice, glare can be reduced by:

- Increasing the mounting height of lights
- Reducing the number of non cut-off lights that are visible by the user
- Using cut-off lights
- Increasing the shielding angles of cut-off lights

1.4.7 Color Aspects

The quality of the color of light emitted by lighting sources is characterized by two properties:

1. Color appearance
2. Color-rendering capabilities

Both properties of a light source are determined by the spectral composition of the light emitted. Different spectral compositions of light can result in a similar color appearance, but they can produce different color rendering.

1.4.8 Color Appearance

The color appearance refers to the apparent color (chromaticity) of the light emitted by a given lighting source and is described by the correlated color temperature (*T*cp) in Kelvin (K). There are three main groups of color appearance according to their correlated color temperature: warm (*T*cp < 3300 K), intermediate (3300 K ≤ *T*cp ≤ 5300 K), and cool (*T*cp > 5300 K). The choice of color appearance depends on the illuminance level, the colors of the illuminated surfaces, and the ambient climate. The color temperature preferred by users increases together with the increase in the illuminance level. This means that lamps with a cool color appearance should be used at high levels of illumination (above 750 lx), while lamps with a warm color appearance should be used at low illuminances (below 300 lx). Surfaces or objects with warm colors, such as yellow or orange, look better and more natural when they are illuminated by lamps with warm light (color appearance). Similarly, objects with cool colors, such as blue, gray, or silver, appear more natural when illuminated by cool light. Cooler artificial light is also preferred in warm climates, where natural light is generally cooler, whereas warmer artificial light is recommended in cold climates.

1.4.9 Color Rendering

Color rendering describes the appearance of object colors under a given light source compared with their appearance under a reference source. A general color-rendering index, R_a, has been introduced for the objective identification of color-rendering properties. Its maximum value of 100 represents excellent color rendering. Perceived color saturation is lower under lamps with poorer color rendering than under lamps having better color rendering.

Generally, in interiors where people work or stay for long periods, lamps should not have an R_a less than 80. The recommended minimum values of *R*a are provided in the standards by type of interior or activity. For visual tasks that demand color recognition, such as color inspection in multicolored printing or working with precious stones, lamps should be used that have color rendering of at least 90. Lamps usually used for high-bay lighting can have a color rendering index that is lower than 80. It must be pointed out that the safety color should be recognizable even when the recommended minimum value is very low ($R_a = 20$).

1.4.10 Flicker

Flicker is the impression of unsteadiness of visual sensation induced by a light stimulus whose luminance or spectral distribution fluctuates over time (CIE

2001). It usually causes distraction and can cause headaches or various visual complaints. Light emission of all AC lamps has a cyclical variation. This is small for filament lamps, but more visible in discharge lamps. Flicker usually increases with the lamp's age (especially for fluorescent lamps) and can be avoided by replacing old lamps with new ones. Flicker from high-pressure lamps is more noticeable in lamps with transparent bulbs than in those with fluorescent coating on the bulb.

Moving parts of rotating or reciprocating machinery, which are illuminated by discharge lamps, can appear to be stationary, operating at a reduced speed, or even in reverse rotation. This phenomenon is called the stroboscopic effect, and it may be a cause of occupational accidents. Flicker and stroboscopic effects should be avoided by the design of an appropriate lighting system. The use of high-frequency electronic control gear for discharge lamps can more effectively reduce flicker and stroboscopic effects than conventional methods, such as dividing lamps between the three phases of the electricity supply or using anti-stroboscopic systems in lights.

1.5 CONCLUSIONS

The objective of illumination engineering is not only to avoid any possible danger or other negative results related to lighting but also to actually promote health as defined by the World Health Organization: "Health is a state of complete physical, mental, and social well-being and not merely the absence of disease or infirmity" (http://www. who.int/about/definition/en/). This means that lighting should ensure a human-friendly, luminous environment that is appropriate to the visual tasks performed. Therefore, visual strain should be minimized and well-being ensured during visual work under a given lighting system. Longitudinal studies of lighting in offices (Çakir 1991) have shown that artificial lighting is far from promoting health and well-being in the spirit of the WHO's definition of health, and technical improvements in office illumination would not, over the years, lead to an increase in the level of acceptance of artificial lighting. The variety of lighting implementation for a given interior or workplace, while ensuring compliance with lighting requirements, gives lighting designers an opportunity to create a wide range of luminous environments. Even so, this does not mean that they all affect the well-being of users in the same way. If users feel uncomfortable under a given lighting system, they can suffer from various visual disorders. Differences are related to age and gender regarding lighting preferences.

In general, women prefer warmer light and less bright light on surfaces in the visual field than men. Furthermore, the preference for using additional local lighting increases with the age of the users (Wolska 2003). From the point of view of the human aspects in lighting, it is important for lighting designers to choose the best lighting systems while considering the preferences of users.

REFERENCES

Boyce, P. R. (2003). *Human factors in lighting* (2nd ed.). Taylor & Francis.

Çakır, A. E. (1991). *An investigation on state-of-the-art and future prospects of lighting technology in German office environments.* Ergonomic Institute for Social and Occupational Sciences Research.

European Committee for Standardization (CEN). (2002a). *Light and lighting: Lighting of work places. Part 1: Indoor work places* (Standard No. EN-12464-1:2002). CEN.

European Committee for Standardization (CEN). (2002b). *Light and lighting: Basic terms and criteria for specifying lighting requirements* (Standard No. EN 12665:2002). CEN.

International Commission on Illumination (CIE). (2001). *Lighting of indoor work places* (Standard No. CIE S 008/E-2001). CIE.

International Organization for Standardization (ISO). (2002). *The lighting of indoor work places* (Standard No. ISO/CIE 8995:2002). ISO.

Rea, M. S. (1993). *Lighting handbook: Reference & application* (8th ed.). Illuminating Engineering Society of North America.

Wolska, A. (2003). Visual strain and lighting preferences of VDT users under different lighting systems. *Journal of Safety and Ergonomics (JOSE)*, *9*(4), 431–440.

FURTHER READING

International Commission on Illumination (CIE). (2020). *CIE S 017/E:2020 ILV: International lighting vocabulary* (2nd ed.). CIE.

International Commission on Illumination (CIE). (2005). *CIE S 015/E:2005 Lighting of outdoor work places.* CIE.

International Organization for Standardization (ISO). (2018). *ISO/CIE 8995-3:2018(E) lighting of work places – Part 3: Lighting requirements for safety and security of outdoor work places.* ISO.

International Organization for Standardization (ISO). (2004). *ISO 15469:2004(E)/CIE S 011/E:2003 spatial distribution of daylight - CIE standard general sky.* ISO.

International Organization for Standardization (ISO). (2025). *ISO/CIE 8995-1:2025(en) light and lighting - Lighting of work places - Part 1: Indoor.* ISO.

Illumination
Basic Definitions

2

P. R. Boyce

2.1 PHOTOMETRIC QUANTITIES

Illumination is the process of placing light on an object. Light is electromagnetic radiation in the wavelength range 380 to 780 nm. Radiation in these wavelengths is absorbed by the photoreceptors of the human visual system, which initiates the process of seeing.

The total electromagnetic radiation emitted by a source is quantified as the radiant flux. Radiant flux is a rate of flow of energy and is measured in watts. The total amount of light emitted by a source is quantified as the luminous flux. Luminous flux is radiant flux multiplied by the relative spectral sensitivity of the human visual system, wavelength by wavelength. Luminous flux is measured in lumens. The relative spectral sensitivity conventionally used is the standard photopic observer of the Commission Internationale de l'Eclairage (CIE). Luminous flux is used to quantify the total light output of a light source in all directions.

To quantify the light emitted in a given direction, luminous intensity is used. Luminous intensity is the luminous flux emitted/unit solid angle, in a specified direction. The unit of measurement is the candela, which is equivalent

DOI: 10.1201/9781003675761-2

TABLE 2.1 The Photometric Quantities

MEASURE	*DEFINITION*	*UNITS*
Luminous flux	That quantity of radiant flux which expresses its capacity to produce visual sensation	Lumens (lm)
Luminous intensity	The luminous flux emitted in a very narrow cone containing the given direction divided by the solid angle of the cone, i.e. luminous flux/unit solid angle	Candela (cd)
Illuminance	The luminous flux/unit area at a point on a surface	Lumens/m2
Luminance	The luminous flux emitted in a given direction divided by the product of the projected area of the source element perpendicular to the direction and the solid angle containing that direction, i.e. luminous flux/unit solid angle/unit area that direction, i.e. luminous flux/unit solid angle/ unit area	Candela/m2
Reflectance	The ratio of the luminous flux reflected from a surface to the luminous flux incident on it	
For a matte surface	Luminance = (illuminance * reflectance)/π	

to a lumen/steradian. Luminous intensity is widely used to quantify the distribution of light from a luminaire.

Both luminous flux and luminous intensity have area measures associated with them. The luminous flux falling on the unit area of a surface is called the illuminance and is measured in lumens/m^2 or lux. The luminous intensity emitted per unit projected area in a given direction is the luminance and is measured in candelas/m^2. Table 2.1 summarizes these photometric quantities.

2.2 COLORIMETRIC QUANTITIES

The photometric quantities do not differentiate between light made up of different wavelength combinations and hence that may appear different in color. The characteristic of light that largely determines what color is perceived is the spectral power distribution of the light reaching the eye, i.e. the amount of radiant flux present in each wavelength interval. The CIE has developed two three-dimensional color spaces, Lab and Luv, both based on mathematical manipulations applied to spectral power distributions. Lab is used for object colors and Luv for self-luminous colors. If two colors have the same coordinates in either of these color spaces, the two colors will appear the same. The distance two colors are apart in color space is related to how easily they can be distinguished.

Although the Luv and Lab color spaces are the most sophisticated characterizations of color, other simpler color systems are used by the lighting industry. The CIE 1964 uniform color space is used in the calculation of the CIE General Color Rendering Index, an index widely used to classify the color rendering capabilities of light sources. The positions in the color space of eight test colors under a reference light source and under the light source of interest are calculated. The separation between the two positions of each test color are obtained, and the separations are summed and scaled to give a value of 100 when there are no separations for any of the test colors, i.e. for perfect color rendering. This is a very crude system. Different light sources have different reference light sources, and the summation means that light sources that have the same sum of separations with different combinations of the test colors can have the same color rendering index.

A two-dimensional color surface is used to characterize the color appearance of light sources and to define the acceptable color characteristics of light signals. This is the CIE 1931 chromaticity diagram (Figure 2.1). This diagram is a slice through color space at a fixed luminance. The curved boundary of the surface represents colors created from single wavelengths and is called the spectrum locus. The point at the center, called the equal energy point, is where a spectrally neutral surface will be located. Any other color will be located somewhere between the equal energy point and the spectrum locus. The location of a specific color can be quantified by its chromaticity coordinates or by two measures, dominant wavelength and excitation purity. To determine the dominant wavelength of a surface lit by a known light source, a line is drawn through the chromaticity coordinates of the light source alone and the surface when lit by the light source, and extended to the spectrum locus. The wavelength at which the extended line intersects the spectrum locus is the dominant wavelength. As for excitation purity, this is the ratio of the distance from the chromaticity coordinates of the light source to those of the lit surface, divided by the total distance from the coordinates of the light source to the intersection of the line with the spectrum locus.

The other notable feature of Figure 2.1 is the curve through the middle of the diagram. This is the Planckian locus, representing the chromaticity coordinates of a full radiator at different temperatures. The color appearance of a nominally white light source is conventionally described by its correlated color temperature. This is the temperature, in degrees Kelvin, of the full radiator that is closest to the light source on the CIE 1931 chromaticity diagram.

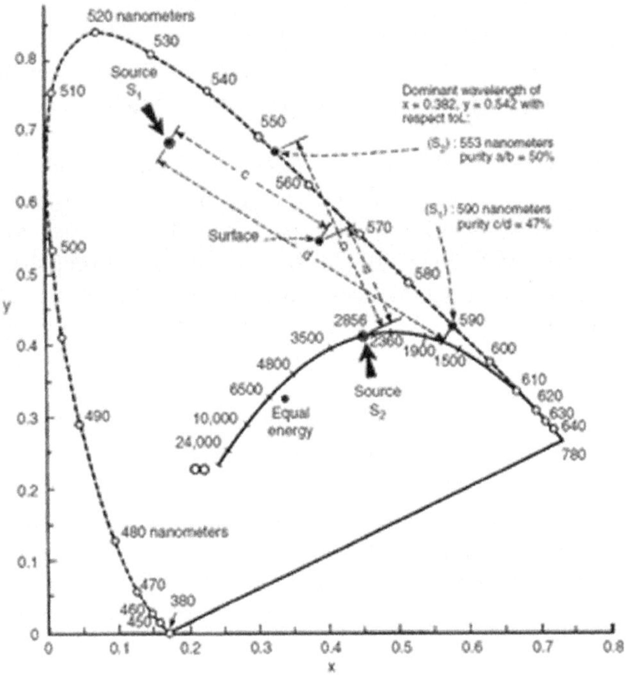

FIGURE 2.1 The CIE 1931 chromaticity diagram showing the spectrum locus, the Planckian locus, the equal energy point, and the method of calculating dominant wavelength and excitation purity for a surface and two different light sources (from IESNA 2000).

2.3 THE GENERATION OF LIGHT

Light occurs naturally, in the form of daylight, and is created deliberately, usually by converting other forms of energy into electricity and then into light. The most obvious characteristic of natural light is its variability. Daylight varies in magnitude, spectral content, and distribution with different meteorological conditions, at different times of day and year, and at different latitudes. Daylight can be divided into two components, sunlight and skylight. Sunlight is light received directly from the sun and produces strong, sharp-edged shadows. Skylight is light from the sun scattered in the atmosphere, and it produces only weak shadows. The balance between sunlight and skylight is determined by the nature of the atmosphere and the distance that the light passes through

TABLE 2.2 Summary of Characteristics of Some Common Electric Light Sources

LIGHT SOURCE	*LUMINOUS EFFICACY (LM/W)*	*CORRELATED COLOR TEMPERATURE (K)*	*CIE GENERAL COLOR RENDERING INDEX*	*LAMP LIFE (H)*	*APPLICATIONS*
Tungsten	8–19	2700	100	750–2000	Domestic retail
Tungsten–halogen	8–20	2900	100	2000–6000	Display
Tubular fluorescent	60–1[illegible]0	3000–5000	50–95	9000–20000	Commercial
Compact fluorescent	50–70	2700–4100	80–85	9000–20000	Commercial retail
Mercury vapor	30–60	3200–7000	15–50	16000–24000	Agricultural industrial
Metal halide	50–1[illegible]0	3000–6500	65–95	3000–20000	Industrial commercial exterior
High pressure sodium	60–140	2100–2500	20–70	10000–24000	Industrial exterior
Low pressure sodium	100–180	1800	NA	16000–18000	Exterior

it. The greater the amount of water vapor and the longer the distance, the higher the proportion of skylight.

Electric light sources used for illumination can be divided into two classes: incandescent lamps and discharge lamps. Incandescent lamps produce light by heating a filament. Discharge lamps produce light by an electric discharge in a gas. Incandescent lamps operate directly from the main electrical supply. Discharge lamps require control gear between the lamp and the electricity supply because different electrical conditions are required to initiate the discharge and to sustain it.

Table 2.2 summarizes the performance characteristics of two incandescent lamp types and six discharge lamp types and gives the most common applications for each lamp type. The values in the Table 2.2 should be treated as indicative only. Details about the characteristics of any specific lamp should be obtained from the manufacturer.

Solid-state light sources, such as light-emitting diodes, are being developed rapidly. At present, they are widely used for signs and signals but have not yet been adopted for illumination.

2.4 CONTROL OF LIGHT OUTPUT

To produce illumination, it is necessary to both generate light and to control its distribution. For daylight, light distribution is controlled by means of aperture shape and placement. For electric light sources, light distribution is controlled by placing the light source in a luminaire. The luminaire provides electrical and mechanical support for the light source and controls the light distribution by means of reflection, refraction, or diffusion, individually or in combination.

As for control of light output, the admission of daylight through a window can be limited by fixed light shelves or adjustable blinds. Wherever the sun, or a very bright sky, is likely to be directly visible through a window, some protection is desirable. The best form of protection is that which shields the viewer from a direct view of the sun or bright sky while preserving some view out.

For electric light sources, control of light output is provided by switching or dimming. Time switches are used to switch off all or parts of a lighting installation at the end of the working day. Occupancy sensors are used to switch off lighting when there is nobody in the space. Such switching systems reduce electricity consumption, but they may be irritating if they switch lighting off when it is required, and they may shorten lamp life if switching occurs excessively. Manual switching is more likely to be used if the lighting being switched is visible from the control panel and the control panel itself is labeled so that the operator knows which luminaires are being switched.

As for dimming systems, these all reduce light output and energy consumption, but a different system is required for each lamp type. The factors to consider when evaluating a dimming system are the range over which dimming can be achieved, without flicker; the extent to which the color properties of the lamp change as the light output is reduced; and any effect dimming has on lamp life.

2.5 THE HUMAN VISUAL SYSTEM

The human visual system involves both eye and brain working together. Light entering the eye is focused on the retina by the combined optical power of the air/cornea surface and the lens of the eye. At the retina, the incident photons of light are absorbed and converted to electrical signals, following which some image processing occurs. The processed images are transmitted up the optic nerves to the visual cortex where they are interpreted in terms of past experience.

There are two types of visual photoreceptors in the retina, called rods and cones. These photoreceptors have different wavelength sensitivities, different absolute sensitivities to light, and are distributed differently across the retina. Rods are approximately 100 times more sensitive than cones and effectively provide a night retina. Cones operate during daytime. Cones are concentrated in a small central area of the retina called the fovea, although there are some cones distributed evenly across the rest of the retina. Rods are absent from the fovea, reaching their maximum concentration about 15 from the fovea. This variation in concentration of rods and cones is amplified by the number of photoreceptors connected to each optic nerve fiber. The ratio of photoreceptors to optic nerve fibers is close to one in the fovea but increases rapidly as the deviation from the fovea increases. This change in the number of photoreceptors connected to each optic nerve implies an increase in receptive field size as deviation from the fovea increases. Functionally, the fovea provides fine discrimination of detail and color, while the periphery is devoted to detecting changes in the visual field that require the attention of the fovea.

The human visual system can operate over a range of about 12 log units of luminance, from starlight to bright sunlight, but it cannot operate over the whole range simultaneously. At any instant, the visual system can cover a range of about 2–3 log units of luminance. Luminances outside this range are seen as either glaringly bright or uniformly black. To move over the complete

TABLE 2.3 Summary of the Operating States of the Visual System

	LUMINANCE RANGE		*PEAK WAVELENGTH*	
OPERATING STATE	*PHOTORECEPTOR*	*ACTIVE*	*SENSITIVITY*	*CHARACTERISTICS*
Photopic	>3 cd/m2	Cones	555 nm	Fine resolution Good color vision
Scotopic	<0.001 cd/m2	Rods	507 nm	No color vision Poor resolution Fovea "blind"
Mesopic	>0.001 and <3 cd/m2	Cones and rods	555 nm in fovea, between 555 and 507 nm elsewhere	Reduced color Reduced resolution relative to photopic

range of luminances, a process of adaptation has to occur. The capabilities of the visual system depend on the luminance to which it is adapted. Three ranges of luminance can be distinguished: photopic, mesopic, and scotopic. Table 2.3 summarizes these ranges and the associated visual capabilities.

Interior lighting and decor usually provide luminances which ensure the visual system is in the photopic state. Exterior lighting on roads and urban areas usually provides luminances high enough to ensure at least mesopic vision. It is only in areas without lighting that the scotopic state is reached.

2.6 TASK PERFORMANCE AND VISUAL PERFORMANCE

Illumination is often provided to facilitate the performance of tasks. Visual tasks usually have three components: visual, cognitive, and motor. The visual component is the process of extracting information relevant to the performance of the task using the sense of sight. The cognitive component is the interpretation of the sensory stimuli. The motor component is the process by which the stimuli are manipulated to extract information and/or the actions decided upon are carried out. Every task is unique in its balance between visual, cognitive, and motor components and hence in the effect illumination has on task performance.

While exact prediction of the relationship between illumination and task performance is not possible without an understanding of the balance between visual, cognitive, and motor components, the form of the relationship for tasks with a large visual component is well established. Figure 2.2 shows the relative visual performance of a task requiring the detection of a stimulus in a known position, for four different size stimuli, for a range of luminance contrasts and retinal illuminances. The overall shape of the relative visual performance surface shows that the visual system is capable of a high level of visual performance over a wide range of visual sizes, luminance contrasts, and retinal illuminations, but at some point either visual size, luminance contrast, or retinal illumination will become insufficient, and visual performance will rapidly collapse toward threshold and invisibility. The existence of a plateau of visual performance implies that for many visual tasks, visual performance is insensitive to illumination, the visual system being sufficiently flexible to cope equally well with a wide variety of visual stimuli.

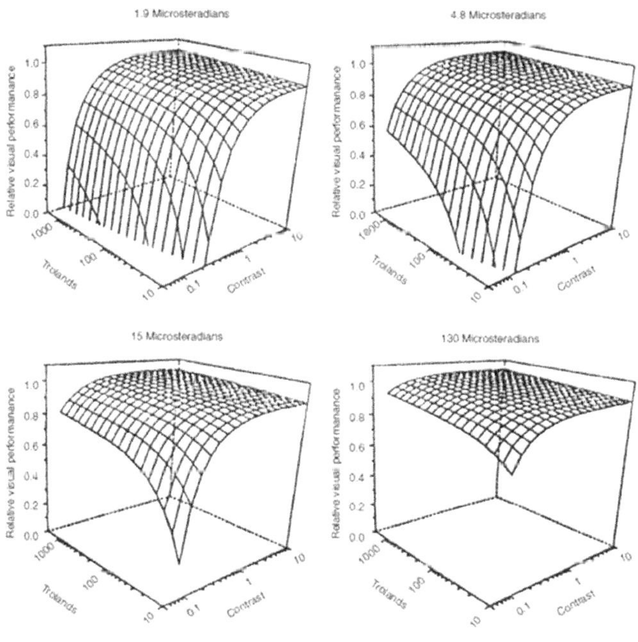

FIGURE 2.2 Four relative visual performance surfaces for stimuli of four different sizes, measured in microsteradians. Each surface shows relative visual performance plotted against retinal illumination in trolands and luminance contrast. Retinal illumination in trolands is calculated as the product of the luminance of the surface being viewed and the associated pupil size (from IESNA 2000).

The relationship shown in Figure 2.2 is derived from tasks that are viewed directly, i.e. that are imaged on the fovea. However, there are a whole class of tasks in which the object to be detected can appear anywhere in the visual field. These tasks involve visual search. Visual search is typically undertaken through a series of eye fixations, the fixation pattern being guided either by expectations about where the object to be seen is most likely to appear or by what part of the visual scene is most important. The speed with which a visual search task is completed depends on the visibility of the object to be found, the presence of other objects in the search area, and the extent to which the object to be found is different from the other objects. The most difficult visual search task is one where the object to be found is of low visibility, is situated in a cluttered field, and where the clutter is very similar to the object to be found, e.g. searching for a face in a crowd.

The level of visual performance that is achievable with a given task depends on the stimulus the task presents to the visual system and the operating state of the visual system. There are five characteristics of the task and its lighting which define it as a stimulus to the visual system. These are its visual size, luminance contrast, color difference, retinal image quality, and retinal illumination.

2.6.1 Visual Size

The larger the stimulus, the easier it is to detect. There are several different ways to express the size of a stimulus presented to the visual system, but all of them are angular measures. The visual size of a stimulus for detection is best given by the solid angle the stimulus subtends at the eye. The visual size for resolution is usually given as the angle the critical dimension of the stimulus subtends at the eye. For complex stimuli, the measure used to express its dimensions is the spatial frequency distribution. Lighting can change the visual size of three-dimensional stimuli by casting shadows which extend the visual size of the stimulus.

2.6.2 Luminance Contrast

The higher the luminance contrast of a stimulus, the easier it is to detect. There are two different forms of luminance contrast. For stimuli which are seen against a uniform background, e.g. print, luminance contrast is defined as

$$C = \frac{(L_t - L_b)}{L_b}$$

where C is the luminance contrast; L_b is the luminance of the background, and L_t is the luminance of the detail.

For stimuli which have a periodic pattern, e.g. a grating, the luminance contrast or modulation is given by

$$C = \frac{(L_{max} - L_{min})}{(L_{max} + L_{min})}$$

where C is the luminance contrast; L_{max} is the maximum luminance; and L_{min} is the minimum luminance.

Lighting can change the luminance contrast of a stimulus by producing disability glare in the eye or veiling reflections from the stimulus or by changing the incident spectral radiation where colored stimuli are involved.

2.6.3 Chromatic Difference

It is possible to have a stimulus with zero luminance contrast which can still be detected because it differs from its background in color, i.e. it has a chromatic difference. Chromatic difference can be expressed in terms of positions on an equal-luminance plane of a color space. Light sources with different spectral emissions can alter chromatic differences.

2.6.4 Retinal Image Quality

The quality of the retinal image is determined by the stimulus itself, the extent to which media through which it is transmitted scatter light, and the ability of the visual system to focus the image on the retina. Lighting can do little to alter any of these factors apart from the last. This can be done by increasing the depth of field by using a light source with a spectral power distribution that produces a smaller pupil size.

2.6.5 Retinal Illumination

The retinal illumination determines the state of adaptation of the visual system and the size of the retinal receptive fields, and therefore alters visual capabilities. The retinal illumination is determined by the luminances in the visual field, the pupil size, and the absorption of light in the eye. Illumination and surface reflectances determine the luminances in the visual field.

By altering any or all of these characteristics, the achievable level of visual performance can be moved to the plateau. How effective making any of these changes will be depends on the initial condition. The closer the task is to threshold, the bigger the effect of changing any of the characteristics will be. Above threshold, the most effective characteristic to change is the one which is limiting performance. For example, increasing the luminance contrast of a stimulus that is very small in size will help, but increasing the visual size will be more effective. Similarly, enhancing color difference will have little effect if the task has a high luminance contrast unless the color can be used to distinguish the stimulus from other nonstimulus objects. Often it is not possible to change the characteristics of the task. In this situation, only the lighting can be changed, but it is always desirable to match the lighting to the physical characteristics of the task. For example, if the task is two-dimensional and of matte reflectance located on a matte background, increasing the retinal illumination is about the only option. However, if the object is three-dimensional and is a specular reflector, the appropriate light distribution can increase the visual size by casting shadows, and increase the luminance contrast by producing highlights. If the object is distinguished from its background primarily by color, the light spectrum used is an important consideration.

2.7 VISUAL COMFORT

Ilumination is rarely provided for visual performance alone. Visual comfort is almost always a consideration. The aspects of lighting which cause visual discomfort include those relevant to visual performance and extend beyond them. This is because the factors relevant to visual performance are generally restricted to the task and its immediate area, whereas the factors affecting visual discomfort can occur anywhere within the lit space.

Red, sore, and watering eyes, headaches and migraines, gastro-intestinal problems, and muscular aches and pains can all be due to inappropriate illumination, although all can have other causes. There are several aspects of lighting which can cause discomfort, although whether they do will depend on the context. All can be used to positive effect in some contexts. These aspects are as follows:

2.7.1 Flicker

A lighting installation which produces flicker will be almost universally disliked, unless it is being used for entertainment or in a localized area to attract attention. Lamp flicker can be minimized by using high-frequency control gear for discharge lamps and/or by the mixing of light from lamps powered from different phases of the electricity supply.

2.7.2 Glare

Glare occurs in two ways. First, it is possible to have too much light. This is evident when the observer screws up his eyes, blinks, and looks away. Too much light is rare indoors, but is common in full sunlight. Second, glare occurs when the range of luminance in a visual environment is too large. Glare of this sort can have two effects: a reduction in visual performance and a feeling of discomfort. Glare which reduces visual performance is called disability glare. It is due to light scattered in the eye reducing the luminance contrast of the retinal image. Disability glare is rare in interior lighting but is common on roads at night from oncoming headlights and during the day from the sun.

As for discomfort glare, this, by definition, does not cause any shift in visual performance but does cause discomfort. Discomfort glare increases as the luminance and solid angle of the glare source increase, and decreases as the luminance of the background and the deviation from the glare source increase. Most luminaires are designed to limit the discomfort glare likely to occur for common directions of view.

2.7.3 Shadows

Shadows are cast when light coming from a particular direction is intercepted by an opaque object. If the object is big enough, the effect is to reduce the illuminance over a large area. This is typically the problem in industrial lighting where large pieces of machinery cast shadows in adjacent areas. If the object is smaller, the shadow can be cast over a meaningful area which in turn can cause perceptual confusion. However, shadows are also valuable in revealing the form of three-dimensional objects. Shadows can be reduced by increasing the amount of interreflected light in the space or by providing local lighting which can be adjusted in position.

2.7.4 Veiling Reflections

Veiling reflections occur when a source of high luminance, usually a luminaire or a window, is reflected from a specularly reflecting surface, such as a glossy printed page or a computer monitor. The luminance of the reflected image changes the luminance contrast of the printed text or the display. The two factors which determine the magnitude of veiling reflections are the specularity of the material being viewed and the geometry between the observer, the object, and any sources of high luminance. If the object is completely diffusely reflecting, no veiling reflections occur, but if it has a specular reflection component, veiling reflections can occur. The positions where they occur are those where the incident ray corresponding to the reflected ray, which reaches the observer's eye from the object, comes from a source of high luminance. This means that the strength and magnitude of veiling reflections can vary dramatically within a single lighting installation.

Like shadows, veiling reflections can also be used positively, and when they are, they are conventionally called highlights. Display lighting of specularly reflecting objects is all about producing highlights to reveal the specular nature of the surface.

2.8 LIGHT AND AGING

As the visual system ages, the ability to focus at close distances is reduced, the amount of light reaching the retina is reduced, more of the light entering the eye is scattered, and the color of the light is altered by preferential absorption of the short, visible wavelengths. The consequences of these changes with age are reduced visual acuity, reduced contrast sensitivity, reduced color discrimination, increased time taken to adapt to large and sudden changes in adaptation luminance, and increased sensitivity to glare.

Lighting can be used to compensate for these changes, to some extent. Older people benefit from higher illuminances than are needed by young people, but simply providing more light is not enough. The light has to be provided in such a way that both disability and discomfort glare are carefully controlled and veiling reflections are avoided. Where elderly people are likely to be moving from a well-lit area to a dark area, such as from a supermarket to a parking lot at night, a transition zone with a gradually reducing illuminance is desirable.

2.9 THE HUMAN CIRCADIAN SYSTEM

Exposure to light impacts the human circadian system. The human circadian system involves three components: an internal oscillator, an external stimulus to entrain the internal oscillator, and a messenger hormone, melatonin, that carries the internal "time" information to all parts of the body through the bloodstream. The natural light–dark cycle is one of the most potent of the external entraining stimuli.

By varying the amount of light exposure and when it is presented, it is possible to shift the phase of the circadian system. In addition, exposure to bright light at night can have an immediate alerting effect through the suppression of melatonin. The amount of light required to cause phase shifts or an immediate alerting effect is within the range of current lighting practice. This implies illumination can be used to more rapidly adjust to the need to work at times when one would normally be sleeping. Evidence to date suggests that while such adjustments can be demonstrated in the laboratory, they are difficult to achieve in real-world contexts, probably because of the need to control light exposure over 24 hours.

2.10 TISSUE DAMAGE

In addition to stimulating the visual and circadian systems, exposure to light can cause tissue damage through two mechanisms: photochemical and thermal. Photochemical damage is related to the energy absorbed by the tissue within the repair or replacement time of the cells of the tissue. Thermal damage is determined by the magnitude and duration of the temperature rise.

The factors which determine the likelihood of tissue damage are the spectral irradiance incident on the tissue, the spectral sensitivity of the tissue, the time for which the radiation is incident, and, for thermal damage, the area over which the irradiance occurs. Spectral irradiance will be determined by the spectral radiant intensity of the source of optical radiation; the spectral reflectance and/or the spectral transmittance of materials from which the optical radiation is reflected or through which it is transmitted; and the distance from the source of optical radiation. Area is important for thermal tissue damage because the potential for dissipating heat gain is greater for a small than a large area.

The visual system provides an automatic protection from tissue damage in the eye, for all but the highest levels of visible radiation. This is the involuntary aversion response produced when viewing bright light. The response is to blink and look away, thereby reducing the duration of exposure. Of course, this involuntary response only works for sources which have a high visible radiation component, such as the sun. Sources which produce large amounts of ultraviolet and infrared radiation with little visible radiation are particularly dangerous because they do not trigger the aversion response.

The light source with the greatest potential for tissue damage is the sun. The sun produces copious amounts of ultraviolet, visible, and infrared radiation. However, there exist some electric light sources that can be hazardous, some being intended for lighting and others being used as a source of optical radiation for industrial processes. The photobiological safety of light sources can be evaluated using recommended procedures.

FURTHER READING

Boyce, P.R., 2003, *Human Factors in Lighting* (London: Taylor & Francis).

Illuminating Engineering Society of North America, 1996, *ANSI/IESNA RP-27-96 Recommended Practice for Photobiological Safety for Lamps and Lamp Systems* (New York: IESNA).

Illuminating Engineering Society of North America, 2000, *Lighting Handbook* (9th ed., New York: IESNA).

Rea, M.S., & Ouellette, M.J., 1991, Relative visual performance: a basis for application. *Lighting Research and Technology*, 23, 135–144.

Wyszecki, G., & Stiles, W.S., 1982, *Color Science: Concepts and Methods, Quantitative Data and Formulae* (New York: John Wiley and Sons).

Lighting Equipment and Lighting Systems

3

A. Wolska

3.1 LAMPS

Electrical light sources can be divided into two main groups depending on the method of production of the light: incandescent and discharge lamps. Each group has several subgroups of lamps with a range of lamps that differ in construction, wattage, luminous efficacy, color properties, life, spectral characteristics, etc.

3.1.1 Incandescent Lamps

The incandescent lamp group consists of two main sub-groups: normal incandescent (or tungsten) lamps and tungsten halogen (or halogen) lamps.

DOI: 10.1201/9781003675761-3

3.1.1.1 Tungsten Incandescent Lamp

This is the oldest electrical light source that produces light by electrically heating a tungsten filament to a high temperature (incandescence point). Visible and infrared radiation is emitted. The filament is sealed in a glass bulb (clear, opal, frosted, colored, or mirror coated), which is usually filled with an inert gas, such as argon or krypton. The most common types are known as general lighting service (GLS), decorative (e.g. candle lamps), and reflector lamps (e.g. PAR, crown-silvered lamps). A normal incandescent lamp has a low color-correlated temperature of approximately 2700 K, which refers to the warm color appearance. The light is almost yellow. The advantages of an incandescent lamp are:

- Low initial costs
- Immediate full light output
- Continuous spectral emission
- Excellent color rendering ($R_a = 100$)
- Simple operation
- Ease of dimming

The disadvantages are:

- Low luminous efficacy (7–21 lm W^{-1})
- Short life (approximately 1000 h)
- Life sensitive to voltage fluctuation and vibration

Incandescent lamps are mainly used for domestic and display lighting.

3.1.1.2 Tungsten Halogen Lamps

These lamps, like normal tungsten lamps, produce light from an incandescent filament, but a halogen (i.e. iodine, chlorine, bromine) is added to the normal gas filling. The envelope of these lamps is quartz glass, and they are very small compared with normal tungsten lamps. The halogen regenerative cycle prevents evaporated tungsten particles from blackening the glass envelope and ensures a constant light output. Halogen lamps have an increased efficacy of approximately 30% and an extended life of up to 2000 h compared with normal tungsten lamps. These lamps almost perfectly maintain the light output throughout their life. Tungsten halogen lamps for normal lighting purposes have a correlated color temperature of between 2800 and 3200 K, so the emitted light is whiter (compared with normal incandescent lamps) and correspondingly of a cooler color appearance. Apart from visible and infrared radiation, halogen lamps also emit ultraviolet radiation. Because of adverse effect of ultraviolet (UV) on the eyes and skin, special attention should be paid when using them for local lighting. "UV Stop" or "UV Block" halogen lamps with an inbuilt UV filter should be used instead. Furthermore, these lamps have the same advantages and disadvantages as normal tungsten lamps. There are two main types of halogen lamps: mains-voltage types (i.e. double-ended, single-ended, double envelope, and reflector lamps) and low-voltage types (reflector and capsule lamps).

Halogen lamps are used either for domestic, decorative, accent, task, display, and general lighting purposes, or for projectors and vehicle headlamps.

3.1.2 Discharge Lamps

Discharge lamps can be divided into two main groups: low pressure (fluorescent lamps, induction lamps, and low-pressure sodium lamps) and high pressure (HID) (high-pressure sodium, high-pressure mercury, and high-pressure metal halide lamps).

3.1.2.1 Fluorescent Lamps

These are low-pressure mercury discharge lamps in which light is produced predominantly by fluorescent powders (phosphor) activated by the ultraviolet energy of the discharge. The spectral light distribution depends on the mix of phosphors. A wide range of color appearance (depending on the correlated color temperature) and color rendering index is available. The main advantages are:

- High efficiency (up to 104 lm W)
- Long life (up to 12000 h)
- Wide range of correlated color temperatures (2700–6500 K)
- Wide range of color rendering index (R_a = 50–98)
- Relatively low cost

The disadvantages of fluorescent lamps are:

- Reduction in the luminous flux during the lifetime (after 8000 h it will be approximately 70% of the initial value)
- Blackening of the tube wall (especially at its ends)
- Efficacy sensitive to ambient temperatures (especially in exterior applications, when lamp wall temperature is below optimum value)

There is a wide range of fluorescent lamps:

- Tubular fluorescent lamps (standard, retrofit, high frequency, rapid start, miniature, special: disinfection, medical therapy, Wood's, actinic, UVA, etc.)
- Compact fluorescent lamps, reduced in size with a special design and technology (integrated, nonintegrated, with and without an outer envelope)

Integrated compact fluorescent lamps, which are also called energy-saving lamps, have a built-in ballast and starter circuit in a form that can be simply screwed into an E27 or E14 socket. Nonintegrated compact fluorescent lamps are also energy-saving lamps, but they require luminaires with special sockets and appropriate ballast.

Fluorescent lamps are widely used for either interior lighting (offices, schools, hospitals, factories, etc.) or exterior lighting.

3.1.2.2 Induction Lamps

Induction lamps are a new group of low-pressure mercury vapor discharge lamps. These lamps of electrodeless design are based on induction and gas discharge principles. They provide high quality white light (correlated color temperature 2700–4000 K and color rendering over 80). High frequency operation ensures flicker-free operation and no stroboscopic effect. These lamps have a very long life (60000 h) which makes them practical in places where access is difficult and reduces the inconvenience of maintenance and its related costs. Luminous efficacy is between 64 and 73 lm W^{-1})

Induction lamps are used for the illumination of public buildings, shops, industrial interiors, tunnels, signs, residential areas, obstructions, and explosion-proof applications.

3.1.2.3 Low-Pressure Sodium Lamps

Visible radiation is directly produced in these lamps by sodium discharge. However, light is only emitted in the yellow part of the spectrum, close to the maximum sensitivity of the human eye (555 nm). Efficacy is the highest of all lamps types (up to 203 lm W^{-1}), but the monochromatic light results in no color rendering. The average life is up to 6000 h.

Low-pressure sodium lamps are used where color rendering is of minor importance and contrast recognition is most important, e.g. roads, tunnel and security lighting, harbors, and marshaling yards.

3.1.2.4 High-Pressure Sodium Lamps

These radiate energy across the whole of the visible spectrum. This increases the correlated color temperature (1950–2500 K) and improves color rendering (up to $R_a = 83$). Luminous efficacy is high (up to 150 lm W^{-1}), and life is up to 12000 h.

This type of lamp is used for road lighting, especially in city centers and industrial interior lighting.

3.1.2.5 High-Pressure Mercury Lamps

These emit part of the radiation from the discharge in the visible region and part in the ultraviolet region. The fluorescent powder coating of the inner surface of the outer bulb converts the ultraviolet radiation into visible radiation. Luminous efficacy is medium (up to 61 lm W^{-1}), life is up to 15000 h, and color temperature is in the range of 3300 to 4300 K. The color rendering index is up to 72.

Mercury lamps are used for road lighting, industrial lighting, and for some commercial applications.

3.1.2.6 Metal Halide Lamps

These are high-pressure mercury lamps with other metallic elements, introduced in the form of halides, which operate in the arc tube. The mix of elements gives a wide range of efficacy (up to 110 lm W^{-1}), correlated color appearance (3000–5600 K), and color rendering ($R_a = 65–96$).

Metal halide lamps are used mainly in commercial interiors and industry, as well as for color TV lighting in studios.

The characteristic parameters of the above lamps are listed in Table 3.1.

3.2 LUMINAIRES

A luminaire (light fixture or lighting fitting) is a piece of apparatus that controls the distribution of light given off by a lamp or lamps and which includes all the components required for fixing and protecting the lamps and for connecting them to the supply circuit (Yarham 1994).

Luminaires can take many different forms depending on the type and number of lamps, form and place of application, and efficiency or esthetic priorities.

TABLE 3.1 Lamps Parameters: Performance Comparison

LAMP	*POWER (W)*	*LUMINOUS FLUX (LM)*	*LUMINOUS EFFICACY (LM W–1)*	*COLOUR CORRELATED TEMPERATURE (K)*	*COLOUR RENDERING INDEX (R_A)*
Tungsten incandescent	7–1000	20–18800	7–21	2700	100
Tungsten halogen	5–2000	60–48400	12–33	2800–3200	100
Tubular fluorescent	4–215	140–14600	35–104	2700–6500	50–98
Compact fluorescent integrated	6–23	200–1500	33–65	2700–6500	76–82
Compact fluorescent nonintegrated	5–55	250–4800	46–82	2700–6500	80–92
Induction	55–165	3500–12000	64–73	2700–4000	80
High-pressure mercury	50–1000	1100–58500	11–61	3300–4300	31–72
Metal halide	35–2000	2000–220000	68–110	3000–5600	65–96
High-pressure sodium	35–1000	1300–130000	68–150	1950–2500	20–83
Low-pressure sodium	18–180	1770–32500	100–203		

3.2.1 Optical Control

The following main methods of optical control can be used in luminaires: reflection, refraction, diffusion, polarization, and interference. Light emitted by the lamp is directed on to the work plane by different elements of the optical system, such as the reflector, louver, refractors, diffuser, and shades. The design of the optical system affects the luminaire efficiency (light output ratio), light distribution, degree of brightness control (glare index), appearance, and utilization factor.

3.2.1.1 Reflection

This process takes place when the incident light leaves a surface from the incident side without a change in frequency. Reflection may be specular (polished surfaces), spread (rough surfaces), diffuse (matte surfaces), mixed, selective, or nonselective, depending on the kind of surface. This technique of light control is very efficient, and light can be distributed very precisely as a result of different shapes of reflectors (circular, parabolic, or elliptical), shapes and dimensions of the cells in a grid of louvers, or different finishes of the surface (polished, matte, or rough). Reflectors are used in all kinds of projectors, spotlights, and floodlights and very often in office luminaires, together with another optical element, such as a louver. Luminaires with parabolic polished louvers can control light output and brightness precisely in all directions. They have a high efficiency, good glare control, and can be used in VDT offices.

3.2.1.2 Refraction

This process takes place when the direction of light changes during its transmission from one medium to another in which its speed is different. This technique usually uses prisms (reflecting or refracting) and lenses (cylindrical, prismatic, and stepped). These types of light directors are made of glass or plastics. Luminaires with refractors made of a large number of small prisms (called prismatic controllers) are suitable for general office lighting because of their good glare control and efficiency. However, they are not suitable for VDT rooms, because they do not reduce glare sufficiently to prevent reflections in a VDT screen.

3.2.1.3 Diffusion

This process takes place when the incident light is redirected by uniform scattering in all directions (according Lambert's cosine law), primarily in the process of diffuse transmission or diffuse reflection. Luminance of the surface is the same in all directions. A diffuser scatters the light emitted by a lamp before it leaves the luminaire. For example, a luminaire with an opal diffuser attachment does not significantly change the light distribution of an installed lamp but reduces its overall brightness. It is not recommended for open-plan offices, because the average luminance of diffusers is still rather high and constant in all directions. It could be acceptable in small rooms if the luminaire is not visible at viewing angles in normal work positions.

3.2.1.4 Polarization

This process takes place when unpolarized light is oriented in a defined direction. Polarized light is mostly obtained by transmitting light through multiple refractive layers. Polarized light can reduce veiling reflections and reflected glare, so it can be used in VDT rooms.

3.2.2 Luminaire Characteristics

The main characteristics of luminaires can be listed as follows:

- *Mounting position*: luminaires can be mounted in several ways: fixed on the ceiling (surface mounted), recessed into the ceiling, suspended from the ceiling (pendant mounted), free standing, or wall mounted (see Figure 3.1).

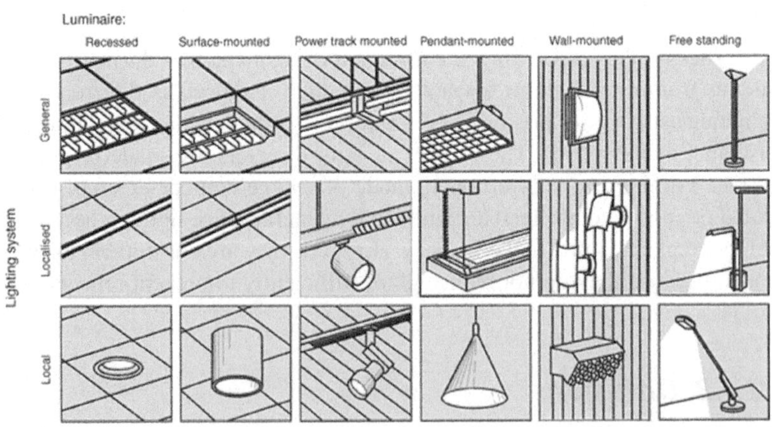

FIGURE 3.1 Luminaire classification by mounting position and lighting system (Dybczynski 1998).

- *Light distribution*: presented graphically by a polar curve shape, which is a schematic illustration of the luminous intensity distribution of the luminaire in characteristic vertical planes. These curves present the way in which the luminaire controls the light from the lamp. Figure 3.2 shows different types of light distribution, from narrow to very broad, and Figure 3.3 shows the classification of luminaires according to the upward and downward direction of luminous flux distribution.
- *Light output ratio*: the ratio of luminous flux emitted by a luminaire to that emitted by lamps used in the luminaire. The luminaire is more efficient the greater the light output ratio and the closer to 1.
- *Utilization factor*: the ratio of luminous flux incident on a work plane to the overall luminous flux emitted by luminaire lamps (all luminaires in an interior). The utilization factor for selected luminaires and their layout in a given interior should be at least 0.5.
- *Operating condition*: the degree of protection against the ingress of dust or moisture is classified according to the ingress protection (IP) system. This system describes a luminaire by a two-digit number, e.g. IP64. The first digit classifies the degree of protection against the ingress of solid bodies, from fingers and tools to dust. The

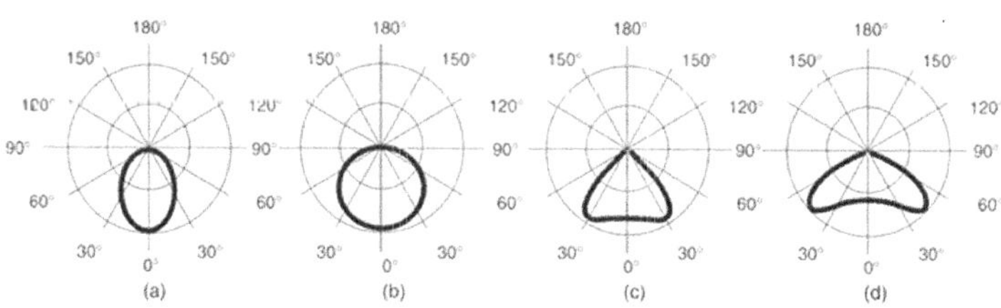

FIGURE 3.2 Typical polar curves: (a) narrow distribution; (b) diffuse distribution; (c) dark–light distribution (usually used in VDT rooms); (d) wide distribution.

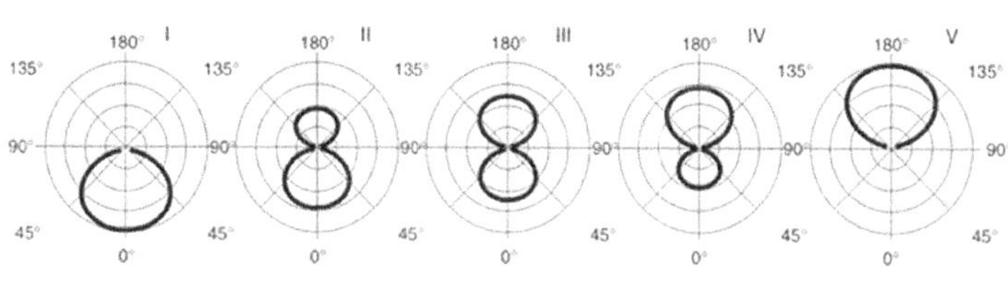

FIGURE 3.3 Classification of luminaires according to the upward and downward direction of luminous flux distribution: I, direct; II, semi-direct; III, diffuse; IV, semi-indirect; V, indirect.

second digit classifies the degree of protection against the ingress of moisture.

- *Electrical protection*: luminaire classes: 0, I, II, and III according to the type of protection provided against electric shocks (where class III indicates the safest usage, the best electrical protection against electric shocks).

3.3 LUMINAIRE SELECTION

The efficiency of the lighting system is a combination of luminaire efficiency and the organization of the work environment, such as task difficulty and location, reflectances of main surfaces internally, and dirt depreciation. Although efficiency is very important, it must be considered together with the luminaire's luminance distribution according to the glare control. Higher efficiencies are often obtained by a wider light output at angles that can cause the glare. Sometimes a chosen luminaire is either efficient or with good glare control but does not integrate esthetically within the environment. Luminaires should be placed into the environment as an integral part of the interior design. The luminaire should also integrate mechanically, electrically, and acoustically with the interior design (Rea 1993).

3.4 LIGHTING SYSTEMS

3.4.1 Classification by Layout of Luminaires

A worker's visual field differs according to whether the worker is concentrating on a task or looking away for relaxation. For this reason, a distinction should be made between task lighting and the lighting of the environment. The following methods of electrical lighting can be distinguished according to the layout of the luminaires in the interior (see Figure 3.1):

- *General lighting*: lighting that provides a uniform level of illumination on the work plane on the whole of an area without any provision for special local requirements.

- *Localized lighting*: lighting that illuminates an interior and at the same time provides a higher illuminance over a particular part or parts of the interior (using luminaires above the visual task).
- *Local lighting*: additional lighting providing illuminance over a small area, for special task requirements, which does not contribute to the general surrounding lighting and is controlled separately from general lighting.

The method of selection of lighting depends on the required illuminance level, room characteristics, and layout of the work stands. General lighting provides the same required illuminance level at each workstation in the room. It is appropriate for small offices (small room, one workstation) or for rooms where the required illuminance is at the same level across the whole of the workstation and task lighting is inappropriate. Localized general lighting provides the required illuminance level at work stands (using task luminaires above the workstation) and a lower level of general illumination. This is desirable in open-plan arrangements. Local lighting provides the required level of illuminance only at that workstation, when an additional task luminaire is mounted. Local lighting should be designed for use only with general lighting because it illuminates only a small area of the workstation. The general lighting level should be at least 20% of the required illuminance level (i.e. 20% of total illuminance from the general and local lighting). This method of lighting is appropriate in large interiors, where work stands differ considerably in illuminance requirements, and work stands are unstable.

3.4.2 Classification by Light Distribution

- *Direct lighting*: lighting by luminaires distributing 90–100% of the emitted light that directly reaches the reference plane (i.e. the surface to be illuminated). This term usually refers to light emitted downwards.
- *Indirect lighting*: lighting by luminaires distributing only 0–10% of the emitted light that directly reaches the reference plane. This means that 90–100% of the emitted light is distributed upwards.
- *Semi-direct lighting*: lighting by luminaires distributing 60–90% of their emitted light downwards and the remaining 40–10% upwards.
- *Semi-indirect lighting*: lighting by luminaires distributing 60–90% of their emitted light upwards and the remaining 40–10% downwards.
- *General diffused lighting*: lighting by luminaires distributing 40–60% of the emitted light downwards and the remaining 60–40% upwards.

A broad range of different possibilities of creating general lighting, such as direct lighting, indirect lighting, and combinations of direct and indirect lighting, gives lighting designers a large number of variants for lighting designs for specific interiors. Choosing one of these lighting systems has its consequences in the appearance of the luminous environment and in energy consumption.

Direct lighting emphasizes horizontal planes, such as the work plane and the floor, as well as vertical surfaces (such as walls), when the luminaires have wider light distribution or when they are placed near the wall. A wide range of direct-type luminaires is available with a variety of light distribution characteristics, from very broad (using translucent diffusing shielding) to narrow concentrated characteristics (using specular reflectors and louvers). Special attention should be paid to glare limitation, especially in VDT rooms. This lighting system has the lowest energy consumption (appropriate illuminance level on the work plane with the lowest energy consumption).

Indirect lighting illuminates the ceiling, so the ceiling is the brightest surface in the visual field. The illumination of the ceiling should be evenly distributed to avoid excessively high values of luminance. Indirect lighting can provide diffuse lighting similar to natural lighting on an overcast day without creating veiling reflections and shadows (the incidence of light from many directions). This lighting system has the highest energy consumption (appropriate illuminance level on the work plane with higher energy consumption).

A combination of direct and indirect lighting can produce excellent results. The indirect portion should not create excessive luminances of the ceiling, and the direct portion should not produce glare.

3.5 MAINTENANCE

Two main factors, namely the reduction of the lamp's light emission over time (lamp aging) and the accumulation of dirt on lamps, luminaires, and other surfaces in the interior, result in a progressively declining level of illuminance over time. Good maintenance of the lighting system means periodic cleaning of luminaires, lamps replaced before the end of their lifetime, and periodic cleaning or painting of the walls and ceiling. The lighting designer should plan appropriate intervals for cleaning, with the frequency depending on the type of luminaire, the rate at which dirt accumulates in a given interior, and the cost of cleaning. There are two types of lamp replacement, called individual and group replacement. Individual replacement (replacement of one or a few lamps in a given interior) is preferred in small interiors because of its high cost. Group replacement (replacement of all lamps in a given interior) is preferred in large interiors and in areas of circulation (i.e. corridors).

3.6 LIGHTING CONTROL SYSTEMS

The demands on modern lighting technology are increasing with time. The aim of contemporary lighting is not only to provide appropriate visual conditions for performing work tasks but also to provide convenience, functionality, and energy conservation. Dimming or brightening the electrical lighting according to daylight changes during the working day is one of the main objectives of lighting control systems. This gives the required level of illuminance from mixed lighting (electrical and daylight) throughout the day and lower costs of energy consumption by the lighting installation. The use of occupancy/motion sensors allows lights to be automatically switched on or off according to occupancy or absence of users in the lighted area/room. Therefore, control systems provide energy and cost savings by reducing illuminance from electrical lighting, as well as by reducing the duration of use. Apart from energy management, a second benefit of the application of lighting control systems is esthetics. Esthetic controls provide the ability to change space functions and can create emotional appeal (Rea 1993). These controls find applications in interiors/areas which are used for more than one purpose and for tasks requiring different lighting conditions (e.g. in conference rooms). Changes of the lighting scene (mood of the illuminated space) can take place rapidly or slowly, depending on the desired effect of excitement or smooth transition.

There are many different lighting control systems from the simplest devices, such as light switches and dimmers, to controls with analogue or digital interfaces. The strategies used for lighting control are manual controls (switches and dimmers), preset control systems (several lighting channels controlled simultaneously in one room), and central control systems (a system for a whole house, which can consist of a central processor and several preset and/or manual controls).

The latest lighting control system installations (preset or central control systems) require designer and electrician knowledge, and training in their operation is usually needed.

REFERENCES

Dybczynski, W., 1998, In Dybczynski, W. and Grzonkowski, J. (eds.) *Oprawy oswietleniowe in Technika Swietlna '98. Poradnik — Informator, tom 1* [*Luminaires in Lighting Engineering '98. Handbook — Information Guide,* Vol. 1], Dybczynski, W., Grzonkowski, J. Eds. (Warsaw: Polski Komitet Oswietleniowy SEP), chap. 9.

Rea, M.S., 1993, *Lighting Handbook. Reference and Application* (8th ed., New York: Illuminating Engineering Society of North America).

Yarham, R., 1994, *Code for Interior Lighting 1994* (London: The Chartered Institution of Building Services Engineers — CIBSE).

FURTHER READING

Belany, P., Hrabovsky, P., & Kolkova, Z. (2021). Combination of lighting retrofit and life cycle cost analysis for energy efficiency improvement in buildings. *Energy Reports*, 7, 2470–2483.

Füchtenhans, M., Grosse, E. H., & Glock, C. H. (2021). Smart lighting systems: State-of-the-art and potential applications in warehouse order picking. *International Journal of Production Research*, 59(12), 3817–3839.

Houser, K. W., Boyce, P. R., Zeitzer, J. M., & Herf, M. (2021). Human-centric lighting: Myth, magic or metaphor?. *Lighting Research & Technology*, 53(2), 97–118.

Olsen, A. A. (2024). Lighting and Illumination. In Olsen, A. A. (ed.) *Applying Physical Ergonomics to Modern Ship Design* (pp. 395–406, Cham: Springer Nature Switzerland).

Tan, Y., Chen, P., Shou, W., & Sadick, A. M. (2022). Digital Twin-driven approach to improving energy efficiency of indoor lighting based on computer vision and dynamic BIM. *Energy and Buildings*, 270, 112271.

4 IoT-Based Solar Streetlight Systems in Improving Human Factors and Ergonomics for Outdoor Environment

Challenges and Prospects

P. L. Chong, P. Manokaran, P. K. Ng, and F. Y. Kong

DOI: 10.1201/9781003675761-4

4.1 INTRODUCTION

The term "Internet of Things," or "IoT," describes a network of networked objects that may share data and communicate with one another over the Internet without the need for human involvement (Chong et al., 2022). These gadgets might be anything from commonplace items like wearable technology and home appliances to business machinery and infrastructure parts. Sensors, actuators, and embedded systems power the Internet of Things, allowing devices to gather, process, and send data (Manoharan et al., 2024; Ganesan et al., 2024). As the device's built-in sensors are linked to an Internet of Things platform, it can integrate data from several devices and apply analytics to share the most useful information for specific applications (Chong et al., 2022). These advanced IoT platforms can determine what information is relevant and what may be safely ignored. The data can be used to identify patterns, make recommendations, and predict potential problems before they arise. The implementation of the IoT revolutionises numerous fields, enhancing efficiency and functionality across diverse applications such as in smart farming (Al-badani et al., 2023a; Chong et al., 2019a), where IoT sensors monitor soil moisture, temperature, and crop health, optimising irrigation and resource allocation; in energy harvesting systems (Chong et al., 2019b; Al-badani et al., 2023b; Chong et al., 2022), which utilise IoT to gather data on energy consumption; in smart manufacturing (Ng et al., 2023), which integrates IoT devices for real-time monitoring of equipment performance, predictive maintenance, and streamlined production processes; in education (Chong et al., 2022; Chong et al., 2022; Chong & Tan, 2020; Chong et al., 2024), where IoT facilitates personalised learning experiences through connected devices and smart classrooms, fostering interactive and adaptive teaching methods; and in streetlight design, where IoT-enabled lighting solutions adjust brightness levels based on traffic patterns, weather conditions, and pedestrian activity, contributing to energy savings and safer urban environments.

The evolution of street lighting from traditional to solar-powered and finally to IoT-enabled solar streetlights represents a significant advancement in efficiency, sustainability, and functionality. Initially, traditional streetlights relied on grid electricity, often leading to high energy consumption and maintenance costs. The introduction of solar streetlights addressed these concerns by harnessing renewable solar energy to power streetlights, reducing dependency on the grid and minimising environmental impact. However, the integration of IoT technology further enhances solar streetlights by enabling remote monitoring, control, and optimisation of lighting systems. IoT sensors gather data on ambient light levels, traffic patterns, and weather conditions, allowing for dynamic adjustments in brightness and operation schedules. This not only optimises energy usage but also enhances safety and visibility in urban environments. Additionally, IoT-enabled solar streetlights offer features such as predictive maintenance alerts, theft detection, and environmental monitoring, contributing to more sustainable and resilient smart city infrastructures. Overall, the evolution from traditional to IoT-enabled solar streetlights signifies a paradigm shift towards smarter, greener, and more efficient urban lighting solutions to provide benefits to the user in the form of improved human factors and ergonomics.

In connection to the above discussion, human factors and ergonomics is an area of study that investigates the interaction of humans with systems, products, and environments, aiming to optimise efficiency, safety, and user experience by considering human capabilities and limitations (Tan et al., 2014; Ng et al., 2016; Tan et al., 2021; Kang et al., 2022; Lim & Ng, 2021). Human factors and ergonomics are crucial because they enhance the design of systems and environments to match human capabilities, ultimately improving efficiency, safety, and overall user experience (Yeow et al., 2014). Smart IoT solar streetlights are pivotal in improving human factors and ergonomics of drivers on the road through their integration of intelligent features and sustainable design. These lights prioritise user safety and comfort by dynamically adjusting brightness levels based on environmental conditions and user presence, enhancing visibility, and minimising accident risks. Leveraging solar energy, they promote sustainability while ensuring reliable lighting and fostering a sense of security in outdoor environments. Their user-centric design, incorporating motion sensors and adaptive lighting, caters to diverse user needs, including those with disabilities, thereby enhancing accessibility. Furthermore, by enabling remote monitoring and maintenance, these streetlights optimise operational efficiency and minimise downtime, ensuring uninterrupted illumination. Through data collection and analysis, they offer insights into usage patterns and environmental factors, informing urban planning decisions for more ergonomic and pedestrian-friendly public spaces. Overall, smart IoT solar streetlights epitomise the synergy between technology, environmental consciousness, and user-centric design, positively impacting human factors and ergonomics in outdoor lighting solutions.

Hence, this book chapter aims to provide a comprehensive review of the various research on IoT solar streetlights, to provide insight into improving human factors and ergonomics of users, and to highlight the challenges and prospects for this technology.

4.2 REVIEW OF VARIOUS IOT SMART STREETLIGHT DESIGNS

Research into improving IoT solar streetlights is crucial due to its potential to enhance urban infrastructure, promote sustainability, and improve safety for pedestrians and drivers alike. Coupled with human factors research, attention to ergonomic design elements such as intuitive control interfaces and ergonomic placement of sensors can not only improve user experience but also enhance pedestrian and driver safety by minimising distractions and optimising interaction with the IoT solar streetlight system. Numerous studies focus on optimising various aspects of IoT solar streetlights, including energy efficiency, performance reliability, smart functionality, and user-centric design.

One significant area of research revolves around enhancing energy harvesting and storage capabilities to ensure consistent operation, especially in regions with limited sunlight or adverse weather conditions. This involves developing more efficient solar panels, advanced battery technologies, and intelligent power management systems to maximise energy utilisation and storage capacity. Furthermore, researchers are exploring advanced sensing technologies and data analytics algorithms to improve the functionality of IoT-enabled streetlights. These include integrating sensors for real-time monitoring of environmental parameters like air quality, temperature, and humidity, as well as traffic and pedestrian flow. Data analytics techniques enable predictive maintenance, adaptive lighting control, and dynamic adjustment of lighting levels based on real-time conditions, enhancing safety and energy efficiency. Additionally, research efforts focus on enhancing connectivity and communication protocols to ensure seamless integration and interoperability of IoT devices within smart city infrastructures. This involves developing robust wireless communication standards, secure data transmission protocols, and cloud-based platforms for centralised monitoring and control of street lighting networks.

To benefit pedestrians and drivers on the road, several improvements have been implemented in IoT solar streetlights. These include adaptive lighting where IoT-enabled streetlights adjust brightness levels based on real-time traffic conditions, improving visibility and safety for drivers and pedestrians while minimising energy waste during low-traffic periods. Besides that, smart intersections capability is also integrated into IoT solar streetlights, as the integration of IoT sensors enables smart intersections that detect approaching vehicles and pedestrians, optimising traffic flow and reducing the risk of accidents. In addition, the IoT solar streetlight is also designed to incorporate emergency response systems. IoT streetlights equipped with sensors can detect emergencies such as accidents or fires and automatically illuminate the affected areas, aiding emergency responders and enhancing safety for pedestrians and drivers. Moreover, IoT solar streetlights are also used to facilitate environmental monitoring. Sensors embedded in IoT solar streetlights can monitor air quality and pollution levels, providing valuable data for urban planners to implement measures to improve environmental health and mitigate health risks for pedestrians and drivers. Thus, ongoing research endeavours aim to continuously improve IoT solar streetlights, addressing energy efficiency, functionality, and safety concerns to create smarter, more sustainable urban environments for pedestrians and drivers alike. The following sections provide details of some of the latest, relevant research on innovating and enhancing the functionality of IoT solar streetlights for the benefit of users.

4.2.1 IoT-Based Smart and Intelligent Smart City Energy Optimisation

An energy-efficient, smart, and intelligent street road lighting system based on an Internet of Things framework is proposed in Chen et al. (2022). It consists of an intelligent electric pole with a controller for adjusting LED lamps and an Internet of Things sensor base. Based on human eye sensitivity, the researchers replaced conventional metal halide lamps with mesopic design LED lamps, which also resulted in significant energy savings. Smart and intelligent decision-making modules compute intensity levels that generate different width pulses using PWM dimming systems, which in turn trigger LED power switches via DALI controllers mounted inside LED light controllers, based on sensor unit information about traffic flow and the presence or absence of occupants. This study uses sustainable power solutions that include PV solar panel units, battery storage systems, and smart electric power networks. The charging battery system employs an MPPT-based dynamic battery charging algorithm. The results reveal that the proposed energy-efficient smart street lighting saves a significant amount of electricity during peak and off-peak hours, not just on highways but also in residential and suburban pedestrian zones. Finally, it will reduce energy usage and CO2 emissions. Based on a comparative examination of suggested systems with existing lighting systems in terms of energy consumption, the experimental results show that the proposed dynamic battery charging algorithm is more effective than the battery storage PV solar panel system performance.

Thus, the technology developed in this research involves the deployment of IoT devices throughout the city to collect real-time data on energy usage, environmental conditions, and other relevant parameters. This data is then analysed using advanced algorithms and artificial intelligence to optimise energy distribution, reduce waste, and improve overall energy efficiency. By creating a network of interconnected smart devices, the system can respond dynamically to changing energy demands and environmental factors, leading to more sustainable energy practices. The results obtained from this study demonstrate significant improvements in energy optimisation within smart cities. By implementing IoT-based solutions, the researchers were able to achieve higher levels of energy efficiency, reduce overall energy consumption, and lower carbon emissions. The intelligent systems developed through this research have the potential to revolutionise the way energy is managed in urban areas, paving the way for more sustainable and environmentally friendly practices.

The integration of IoT devices and sensors allows for real-time monitoring and control of energy systems, leading to a more stable and reliable energy supply for residents. This reliability in energy provision positively impacts human factors by reducing the likelihood of power outages and disruptions, thus enhancing user comfort and satisfaction. Additionally, the optimised energy distribution facilitated by IoT technology can lead to cost savings for users, further improving the economic ergonomics of individuals within smart cities. Moreover, the research outcomes indicate that by creating more sustainable and energy-efficient urban environments through IoT-based solutions, users can benefit from a healthier and cleaner living space. Reduced energy waste and lower carbon emissions contribute to better air quality and environmental conditions, ultimately promoting the well-being and health of city inhabitants. These improvements in environmental ergonomics can have a direct positive impact on the physical and mental health of users within smart cities.

4.2.2 Internet of Things (IoT)-Based, Solar-Powered, Streetlight System with Anti-Vandalisation Mechanism

Ekaette et al. (2020) describe an Internet of Things (IoT)-based solar-powered street lighting system with an anti-vandalisation mechanism that prioritises power conservation, energy efficiency, automation, and intelligence, as well as vandalism monitoring and tracking. The system is a self-contained solar PV system. It automatically turns on and off the streetlight using the light-dependent resistor (LDR) and saves electricity by using an infrared sensor (IR) and a microcontroller to dim and brighten the LED as needed, extending the life of the lighting module. Furthermore, the system includes a vandalism monitoring/tracking mechanism as well as a user interface (UI) through which the streetlight connects with the control station's people and devices via a Wi-Fi module. This is accomplished using the ESP8266MOD Arduino WiFi module, which functions as the main control system. The lighting module is a 32-bit LED array. When the LDR detects darkness, the system automatically turns on the light to full brightness, while during peak hours when traffic is reduced, the light will dim. When the light detects vehicle movement, it will switch on to full brightness until the vehicle passes. When the LDR detects sufficient brightness such as when the sun is available in the daytime, the light is switched off instead. In addition, the system detects the presence of potential vandals at odd hours, beginning at 11 p.m., and activates an alarm system. Besides that, it also warns the person at the control station about a potential theft threat via Wi-Fi. The system may also trace the vandal via a dual wireless network, including Wi-Fi and GSM, using the triangulation technique. The concept presented in this paper will allow researchers and practitioners to develop smart street lighting systems that are efficient in terms of power consumption while also effectively combating the vandalism of streetlights.

Hence, the technology developed in this research involves the deployment of IoT sensors and devices within solar-powered streetlights. These IoT devices enable remote monitoring and control of the streetlights, allowing for automatic adjustments based on ambient light levels and weather conditions. Additionally, the anti-vandalisation mechanism incorporated into the system triggers alerts in case of tampering or damage to the streetlights, ensuring their continuous operation and longevity. The results obtained from this study demonstrate the effectiveness of the IoT-based, solar-powered streetlight system with the anti-vandalisation mechanism. By harnessing solar energy and IoT capabilities, the streetlights showed improved energy efficiency and sustainability. The anti-vandalisation feature contributed to the security and durability of the streetlights, reducing maintenance costs and ensuring uninterrupted lighting in public spaces.

The implementation of IoT sensors and devices in solar-powered streetlights allows for intelligent monitoring and control of lighting conditions. This results in improved visibility and safety for pedestrians and motorists, enhancing the human factors related to comfort and security in outdoor environments. The automatic adjustments based on ambient light levels ensure that users benefit from well-lit surroundings, contributing to a more ergonomic and user-friendly experience.

Furthermore, the anti-vandalisation mechanism incorporated into the streetlight system adds an extra layer of security for users. By detecting and alerting authorities to any tampering or damage to the lights, the system helps maintain a safe and reliable lighting infrastructure in public spaces. This feature not only enhances the physical safety of individuals but also contributes to their peace of mind and overall well-being, improving the human factors associated with security and trust in the environment. Incorporating user-friendly reporting mechanisms and clear signage regarding the anti-vandalisation features further empowers pedestrians and authorities to collaboratively uphold the integrity of the lighting infrastructure, fostering a sense of ownership and community engagement in maintaining a safe and well-lit environment.

4.2.3 IoT-Based Street Lighting Using Dual Axis Solar Tracker and Effective Traffic Management System Using Deep Learning: Bangladesh Context

An intelligent automation system that can determine luminous control operation based on light intensity was developed by Saifuzzaman et al. (2020) The placement of the LDR sensor allows it to distinguish between day and night using a certain intensity value. Additionally, IR sensors take over the operation of the streetlight system. Here, the power source will be a solar cell, with DC serving as a backup. Additionally, the research made it possible to automatically monitor traffic controls using a video system. A camera with automatic text recovery from video data using lip reading and facial expression decoding using deep learning approaches has been developed. The suggested method includes collecting test data, interpreting the picture frame, and producing text containing the specified words. The proposed methodology will structure test data collecting by incorporating all possible face expressions for various words.

Thus, the prototype developed in this study involves the implementation of IoT devices in street lighting systems equipped with dual-axis solar trackers. These trackers enable the streetlights to dynamically adjust their orientation to maximise solar energy absorption, enhancing energy efficiency and sustainability. Additionally, the integration of deep learning algorithms in the traffic management system allows for real-time analysis of traffic patterns, leading to more effective traffic flow control and congestion management. The results obtained from this research demonstrate the efficacy of the IoT-based street lighting system and traffic management solution. By utilising dual-axis solar trackers, the streetlights showed improved energy efficiency and reduced dependency on the grid, contributing to cost savings and environmental sustainability. The incorporation of deep learning algorithms in traffic management enhanced the system's ability to predict and respond to traffic conditions, leading to smoother traffic flow and reduced congestion in urban areas. The implementation of IoT-based street lighting with dual-axis solar trackers allows for optimised energy efficiency and sustainability. By dynamically adjusting the orientation of streetlights to maximise solar energy absorption, users benefit from well-lit streets while reducing energy costs and environmental impact. This improvement in lighting quality enhances the safety and visibility of pedestrians and motorists, contributing to a more ergonomic and user-friendly environment. Furthermore, the incorporation of deep learning algorithms in the traffic management system enhances the efficiency of traffic flow and congestion management. By analysing real-time traffic patterns and making data-driven decisions, the system can optimise traffic signals and routes, leading to smoother transportation experiences for users. This not only improves the overall efficiency of the transportation network but also reduces travel times and enhances the comfort of commuters.

4.2.4 Energy Efficiency in Smart Street Lighting Using Raspberry Pi Based on SCADA and Internet of Things (IoT)

A smart city requires a reliable, sustainable, and customisable electric energy source. However, electrical energy management must be cost-effective to avoid burdening the local government budget. Public Street Lighting (PSL) is one of the infrastructures that utilise a significant amount of electrical energy and are under the authority of the local government. Lighting from PLS that emits too much light when it is not needed is useless. Currently, typical PLS systems are linked with energy waste. Monitors and controls can only operate locally without the capacity to monitor and control remotely; thus, the response to damage or theft can be slow. The solution is to create a smart PSL system and management. A study by Dwiyaniti et al. (2019) shows a real-world proven solution based on a Raspberry Pi, SCADA, and IoT. The smart PSL will function intelligently in line with its surroundings. Lighting and illumination are arranged according to the presence of people and vehicles to avoid over-lighting and glare. SCADA and IoT technologies are used to monitor labour processes and display data in real-time. The result shows that the smart PLS mode is 43% more efficient than conventional PLS modes.

Hence, the technology developed in this study involves the integration of Raspberry Pi, SCADA systems, and IoT devices into street lighting infrastructure. By leveraging these technologies, the smart streetlights can be remotely monitored and controlled based on real-time data on lighting needs, environmental conditions, and energy consumption. This allows for dynamic adjustments in lighting levels, reducing energy waste and improving overall efficiency. The results obtained from this research demonstrate the effectiveness of the energy-efficient smart street lighting system. By implementing Raspberry Pi, SCADA, and IoT technologies, the study achieved significant energy savings by ensuring that streetlights only operate when necessary and at optimal brightness levels. This not only reduces energy consumption and costs but also minimises light pollution, creating a more sustainable and environmentally friendly lighting solution.

One key aspect highlighted in the research is the ability to remotely monitor and control streetlights in real time using IoT devices and SCADA systems. This capability allows for swift responses to lighting needs, environmental conditions, and energy usage, ensuring that the lighting infrastructure is responsive and adaptive to user requirements. By enabling remote monitoring and control, the system enhances user safety and comfort by ensuring well-lit environments and timely adjustments to lighting levels. Moreover, the research emphasises the importance of energy efficiency in smart street lighting. By utilising Raspberry Pi and IoT technology, the system can operate streetlights only when necessary and at optimal brightness levels, leading to significant energy savings. This not only reduces operational costs but also contributes to environmental sustainability by minimising energy waste and light pollution, creating a more user-friendly and eco-conscious lighting solution. Thus, this research helps improve human factors and ergonomics for users by providing a smart street lighting system that is energy-efficient, responsive, and environmentally friendly. The ability to remotely monitor and control lighting operations enhances user safety and convenience, while the focus on energy efficiency contributes to a more sustainable and user-centric urban environment.

4.2.5 The Prototype of Solar-Powered Building Lighting IoT

Tri Pratiwi et al. (2019) create a prototype of a remote gadget that uses TCP/IP and connects to a local network via a web server. The prototype has two features. First, it can toggle a single lamp and turn off and on the lights throughout the facility. It is controlled using a web application that can also be accessed from cell phones and desktops. The trial results demonstrate that the prototype required a 500-second delay before activating the lamp and fan relays via the web server. The prototype developed in this study involves the implementation of IoT devices in building lighting systems powered by solar energy. By utilising solar panels to generate electricity and IoT sensors for monitoring and control, the system can optimise lighting operations based on natural light availability, occupancy patterns, and user preferences. This integration allows for energy savings, improved lighting efficiency, and enhanced user experience. The results obtained from this research demonstrate the effectiveness of the solar-powered building lighting IoT prototype. By combining solar power with IoT technology, the study achieved energy efficiency by reducing reliance on traditional grid electricity. The smart lighting system's ability to adjust lighting levels based on environmental conditions and user needs not only saves energy but also enhances user comfort and convenience within the building environment. This dynamic control enhances user comfort by providing appropriate lighting levels tailored to specific needs and preferences. The research also demonstrates the effectiveness of the solar-powered building lighting IoT prototype in improving human factors and ergonomics. By optimising lighting operations through IoT technology, the system not only enhances energy efficiency but also creates a more user-friendly environment. The ability to customise lighting settings based on user requirements contributes to a comfortable and ergonomic space for building occupants, promoting well-being and productivity. By providing energy-efficient, sustainable, and user-centric lighting solutions, the study contributes to creating environments that prioritise user comfort, well-being, and efficiency.

4.2.6 Smart Energy Public Street Lighting System

Masri et al. (2024) fabricate an IoT-based public street lighting system that employs solar panels to charge batteries for alternative electrical energy sources for the streetlight. The ESP8266 module controls the system, a Wi-Fi module offers a Wi-Fi connection, a relay module provides an electric switch for turning on or off public street lighting lights via the intelligent energy public street lighting application, and solar panels charge the batteries. The system includes LDR sensor monitoring, ultrasonic sensors, and a Node MCU ESP8266 microprocessor. The light sensor module measures light intensity. A smart energy, public street lighting app is also developed to enable officers to monitor the procedure in real-time. This application will track public streetlights and analyse and graph the results.

The technology developed in this study involves the implementation of a smart streetlight system that utilises IoT technology. By integrating sensors, controllers, and communication devices, the system can dynamically adjust lighting levels based on environmental conditions, traffic patterns, and user needs. This intelligent control mechanism allows for energy-efficient operation, reduced light pollution, and enhanced safety in urban environments. The results obtained from this research demonstrate the effectiveness of the smart streetlight system in improving infrastructure and sustainability. By optimising lighting operations through IoT technology, the system can provide better quality lighting while reducing energy consumption and costs. The dynamic control of lighting levels based on real-time data contributes to a more efficient and user-centric street lighting solution, enhancing the overall quality of life for citizens. This intelligent control mechanism not only contributes to a more sustainable lighting solution but also enhances the overall user experience by providing better quality lighting tailored to specific requirements. By optimising lighting operations through IoT technology, the system enhances the urban environment, making it more user-friendly and ergonomic. The dynamic control of lighting levels based on real-time data contributes to a safer, more efficient, and more comfortable urban space, ultimately benefiting the well-being and experience of users.

4.2.7 Street Lamp Intelligent Monitoring System Design Based on NB-IoT Technology

Tian and Wang (2020) developed a plan for streetlamp monitoring systems using narrow-band Internet of Things (NB-IoT) technology. In this approach, the STM32L431 RCT6 chip serves as the primary control chip, while the NB-IoT module serves as the communication module. It is linked to the Huawei Internet of Things management platform, and application software has been created on the application side to remotely monitor streetlights in real time. The experimental results suggest that the system can perform real-time monitoring and intelligent control of streetlamps.

Thus, the technology developed in this study involves the implementation of an intelligent monitoring system that utilises NB-IoT technology for real-time monitoring and control of streetlamps. By integrating sensors, controllers, and communication devices, the system enables remote monitoring and intelligent control of street lighting operations. This allows for dynamic adjustments in lighting levels based on environmental conditions, traffic patterns, and user needs, ultimately enhancing energy efficiency and user comfort. The results obtained from this research demonstrate the effectiveness of the Street Lamp Intelligent Monitoring System in improving human factors and ergonomics.

By leveraging NB-IoT technology, the system enables real-time monitoring and intelligent control of streetlamps, leading to optimised lighting operations and enhanced user experience. The dynamic control of lighting levels based on data analysis contributes to energy savings, reduced light pollution, and improved safety in urban environments, ultimately benefiting the well-being and satisfaction of users. The user interface of this system prioritises simplicity and intuitiveness, ensuring that operators can easily navigate and utilise its features without extensive training, thereby reducing cognitive load and enhancing overall usability for city officials and maintenance personnel.

4.2.8 A Multifunctional Public Lighting Infrastructure, Design, and Experimental Test

Beccali et al. (2017) describe an experimental multifunctional infrastructure for public lighting that was established in Palermo. The system can provide smart lighting with extra features such as hosting hotspot Wi-Fi, enabling video surveillance, controlling car and pedestrian access, monitoring parking, and supporting environmental monitoring. A remote control and monitoring platform known as "Centro Servizi" processes information from various installations and their status in real-time, sending directives to the devices that govern the luminous flux. Data can be reported either on the web or via a custom app. The study demonstrated the effectiveness of such new infrastructure and its potential to give new services and benefits to citizens, tourists, and public administration. Thus, this system serves as a foundation for the development of many other lighting infrastructure characteristics associated with smart city development.

The prototype developed in this study involves the design and implementation of a multifunctional public lighting system that combines energy-efficient lighting sources with ICT components. By integrating advanced lighting technologies with smart ICT features, the system aims to optimise energy consumption, enhance visual comfort, and provide economic advantages for urban lighting infrastructure. The experimental testing phase evaluates the performance and effectiveness of this integrated lighting solution. The results obtained from this research demonstrate the potential of the multifunctional public lighting infrastructure in achieving its objectives.

Through the integration of efficient lighting sources and ICT, the system shows improvements in energy efficiency, cost-effectiveness, and user comfort. The experimental tests validate the effectiveness of the integrated system in meeting the lighting requirements of urban environments while ensuring sustainability and economic benefits. In short, while optimising energy consumption and enhancing visual comfort, the multifunctional public lighting system prototype developed in this study emphasises human factors and ergonomics in its design and implementation to ensure user satisfaction and operational efficiency.

4.2.9 IoT-Based Automatic Streetlight Control and Fault Detection

Aarthi and Rajeswari (2022) designed a system that can automatically control and identify faulty streetlights. When a streetlight fails to function, the planned fault detector system alerts the local electrical board substation to the problem. Sensors can be installed in the light pole to transmit a signal to the substation if the light does not function properly. The sensor delivers location information to the electricity substation. This allows the operators to simply determine the position from the substation and take the appropriate actions promptly. Sensors are installed to determine whether the fault is with the solar light panel/bulb or the electrical distribution infrastructure. This strategy enables the operator to work and correct it.

Thus, the research product developed in this study involves the integration of IoT devices in street lighting systems to enable automatic control and fault detection. By leveraging IoT sensors and monitoring systems, the streetlights can be remotely controlled based on environmental conditions and user requirements. Additionally, the fault detection mechanism allows for the automatic identification of faulty streetlights, enabling prompt maintenance and ensuring continuous lighting functionality. The results obtained from this research demonstrate the effectiveness of the IoT-based automatic streetlight control and fault detection system.

By automating the control of streetlights and implementing fault detection capabilities, the study achieved improved operational efficiency and reduced downtime due to faulty lights. This not only enhances the overall reliability of the street lighting infrastructure but also contributes to energy savings and improved user experience in urban environments. Incorporating IoT devices into street lighting systems with a focus on ergonomics ensures that the control interfaces are user-friendly and accessible, reducing cognitive load for maintenance personnel and enhancing overall operational efficiency and user satisfaction.

4.2.10 Solar-Wind Power Generation System for Street Lighting Using the Internet of Things

Hossain et al. (2022) designed a hybrid system of solar and wind energy-powered IoT streetlights that can create 12V DC while using no AC converters, lowering the system's cost. The control system was built using IoT technology and incorporated the most sophisticated battery charging mechanism available to extend the life of the battery cells. The hardware system was simulated using EasyEDA and included in the PCB design. The prototype is built with collected data to compare with the theoretical foundation for net-zero energy street lighting (ESL). The prototype resulted in nZESL, and the system's backup stability is ten hours per day, in addition to the validation of theoretical studies and system effectiveness. The system has the potential to contribute significantly to the reduction of CO2 emissions and government subsidies for street lighting.

Hence, the module designed in this study involves the integration of solar and wind power sources with IoT devices for street lighting applications. By combining renewable energy sources and IoT technology, the system can efficiently generate and manage power for streetlights. This approach not only reduces dependency on traditional grid electricity but also promotes environmentally friendly and sustainable lighting solutions. The results obtained from this research demonstrate the effectiveness of the solar-wind power generation system in improving human factors and ergonomics. By utilising renewable energy sources and IoT technology, the system enhances energy efficiency, leading to cost savings and reduced environmental impact. Additionally, the reliable and sustainable power supply contributes to improved lighting quality, safety, and comfort for users, thereby enhancing the overall user experience in urban environments.

4.2.11 System for Providing Flood and Rain Alert (Flood Alert Solar Lamp)

The system developed in Chong and Kong (2022) comprised a solar IoT streetlight with flood detection and rain detection capability. A raindrop sensor is set up to detect the intensity of raindrops in the monitored area and provide a second input signal equal to the intensity of rainfall detected therein. A flood sensor is set up to detect the water level in a monitored surrounding area and provide a first input signal equivalent to the water level detected. Subsequently, an LED is set up to provide a visual signal to the region under observation by lighting, blinking, flashing, or pulsating. In addition, a siren is set up to sound a warning in the monitored environment. For the central control unit, a microcontroller receives the first input signal from the flood sensor and the second input signal from the raindrop sensor.

To create a triggering signal, the microcontroller compares the first and second input signals to respective predefined water level thresholds and predefined raindrop intensity thresholds. Subsequently, a triggering signal is sent to activate the siren and LED when flood and rain are detected. To enable users to take early action for flood evacuation, a global system for mobile communication (GSM) unit is set up to receive the triggering signal from the microcontroller and to generate an alert message warning the users in the case of flooding. A solar panel on a power source management unit is set up to collect solar energy and use it to power the system via a power management circuit. Hence, this technology enhances human factors and ergonomics by providing real-time flood and rain alerts, enabling timely evacuation, and reducing risks, thereby promoting safety and comfort for individuals in flood-prone areas.

4.3 CHALLENGES

From the various research presented in the previous section, it is observable that designing and implementing an IoT-based solar streetlight system to improve human factors and ergonomics for outdoor environments still present several challenges or research gaps that need to be addressed, such as the impact of environmental factors, energy efficiency issue, lighting design, sensor integration issue, communication reliability, maintenance and upkeep of the product, lack of user interface and experience in product, as well as security and privacy of the designed IoT solar streetlight.

The first challenge to be considered is the outdoor environmental factors influencing the streetlight. Outdoor environments are subject to various weather conditions such as rain, snow, extreme temperatures, and humidity. Designing the system to withstand these conditions without compromising functionality and durability is crucial. Next is the consideration of the product's energy efficiency. Solar streetlights rely on renewable energy sources, primarily solar power. Ensuring that the system is optimised for energy efficiency and can harness, and store solar energy effectively is essential for reliable operation, especially during periods of low sunlight. In addition, consideration of the lighting design is also a vital aspect. Proper lighting design is critical for outdoor environments to ensure safety and visibility while minimising glare and light pollution. Designing the system to provide adequate illumination without causing discomfort to pedestrians, cyclists, or drivers requires careful consideration.

Moreover, the challenges of sensor integration to enable accurate, reliable, and consistent sensing is also a dilemma to be solved. Integrating IoT sensors into the streetlight system for functions such as motion detection, ambient light sensing, and remote monitoring adds complexity. Ensuring seamless integration of these sensors with the lighting system and utilising the data collected effectively to optimise performance and energy usage is a challenge. Besides that, IoT-based systems rely on communication networks to transmit data and receive commands. Ensuring reliable communication between individual streetlights, central control systems, and other IoT devices in outdoor environments can be challenging due to factors such as signal interference, network congestion, and connectivity issues.

Other than that, outdoor environments are often more challenging to access for maintenance and repairs. Designing the system with modular components that are easy to replace and incorporating predictive maintenance techniques can help minimise downtime and ensure optimal performance over time. Reducing downtime and optimising performance systems promotes a more predictable and stable work environment for people, thus reducing stress and anxiety associated with unexpected failures or disruptions and enhancing overall user satisfaction and morale. Furthermore, incorporating modular components that are easily replaceable reduces the physical and cognitive workload for maintenance, improving efficiency and reducing the likelihood of errors during maintenance tasks, ultimately enhancing safety and productivity.

The next challenge is in designing the user interface to provide a seamless experience to the operators in interacting with the IoT solar streetlight. Providing a user-friendly interface for configuring and monitoring the streetlight system is essential for operators and administrators. Designing intuitive interfaces that provide real-time insights, alerts, and control options while considering the outdoor environment's usability challenges (such as glare, sunlight, and weather conditions) requires careful attention to human factors and ergonomics principles. For instance, apart from obscuring the visibility of information displayed on screens or interfaces, glare can also cause eye strain, fatigue, and decreased readability, affecting the user's decision-making capabilities, comfort, and safety.

In addition, there is also the consideration of the security and privacy of the system. IoT devices are vulnerable to cybersecurity threats, and ensuring the security and privacy of data transmitted and stored by the streetlight system is paramount. Implementing robust security measures such as encryption, authentication, and access control to protect against unauthorised access and data breaches is essential.

Addressing these challenges requires a multidisciplinary approach involving expertise in electrical engineering, renewable energy systems, lighting design, sensor technology, communication protocols, user interface design, and cybersecurity.

4.4 PROSPECTS

The possible prospects and future works in designing and implementing IoT-based solar streetlight systems to improve human factors and ergonomics in outdoor environments are promising. The prospects can encompass several key areas which are the implementation of advanced sensor technologies, machine learning and artificial intelligence integration, efficient energy storage and management system, a robust communication infrastructure, enabling user-centric design and accessibility, adapting data analytics and visualisation into the system, incorporating community engagement and stakeholder collaboration to further improve the system infrastructure as well as considering product design of the IoT solar streetlight with resilience and climate adaptation design.

Future developments in sensor technologies, such as improved motion sensors, advanced ambient light sensors, and environmental sensors, can enhance the capabilities of IoT-based streetlight systems. These sensors can enable more precise control of lighting levels based on real-time conditions and user needs, further optimising energy efficiency and enhancing user comfort and safety (Chong et al., 2022). In addition, integration of machine learning and artificial intelligence algorithms can enable predictive and adaptive control of streetlight systems. By analysing historical data, weather forecasts, traffic patterns, and user behaviour, AI algorithms can optimise lighting schedules and levels dynamically, improving energy efficiency while meeting the evolving needs of outdoor environments and users (Krishna et al., 2024). Besides that, the continued advancements in energy storage technologies, such as batteries and supercapacitors, can enhance the reliability and autonomy of solar streetlight systems. Future works in energy management techniques, such as demand-response strategies and smart grid integration, can further optimise energy usage and grid interactions, contributing to sustainability and resilience.

Developments in communication technologies, including 5G networks, low-power wide-area networks (LPWANs), and mesh networking protocols, can improve the reliability and scalability of communication infrastructure for IoT-based streetlight systems. Enhanced communication capabilities can support real-time monitoring, remote management, and seamless integration with other smart city applications. Moreover, future works should focus on designing streetlight systems with a strong emphasis on user-centric design and accessibility. This involves considering the diverse needs and preferences of users, including pedestrians, cyclists, motorists, and individuals with disabilities, to ensure that lighting levels, colours, and placement meet universal design principles and enhance overall user experience and safety. Next, leveraging big data analytics and visualisation techniques can provide valuable insights into the performance, usage patterns, and environmental impact of IoT-based streetlight systems. By analysing data collected from sensors, energy meters, and user feedback, stakeholders can identify optimisation opportunities, prioritise maintenance efforts, and make data-driven decisions to continuously improve system efficiency and effectiveness (Sannasy et al.,

2024). Engaging with local communities, stakeholders, and end-users throughout the design, implementation, and operation phases is crucial for the success of IoT-based streetlight projects. Future works should focus on fostering collaboration, transparency, and co-creation processes to ensure that streetlight systems align with community needs, preferences, and sustainability goals. Lastly, with the increasing frequency and intensity of extreme weather events due to climate change, future works should prioritise resilience and climate adaptation strategies in streetlight system design. This includes incorporating robust materials, modular designs, and backup power solutions or rechargeable energy storage stations to withstand environmental hazards and ensure continuity of service during emergencies (Basir et al., 2024).

By addressing these prospects and future works, designers, engineers, policymakers, and stakeholders can collaboratively advance the development and implementation of IoT-based solar streetlight systems that effectively improve human factors and ergonomics in outdoor environments while contributing to sustainability, resilience, and quality of life in urban areas. Collaboration between various stakeholders, including designers, engineers, urban planners, and end-users, is crucial to developing IoT-based solar streetlight systems that effectively improve human factors and ergonomics in outdoor environments.

4.5 CONCLUSION

In conclusion, the development and implementation of IoT-based solar streetlight systems offer significant potential to enhance human factors and ergonomics in outdoor environments. In this chapter, we have explored the challenges and prospects associated with this innovative technology. Challenges such as environmental factors, energy efficiency, sensor integration, communication reliability, maintenance, and user interface design require careful consideration and multidisciplinary collaboration to overcome. However, addressing these challenges presents an opportunity to create more sustainable, efficient, and user-friendly lighting solutions for urban and rural areas alike. Looking ahead, the prospects for IoT-based solar streetlight systems are promising. Advances in sensor technologies, artificial intelligence, energy storage, communication infrastructure, user-centric design, data analytics, and stakeholder collaboration can further enhance the performance, usability, and resilience of these systems. By embracing these opportunities and continuing to innovate, we can create outdoor lighting solutions that not only improve safety and comfort but also contribute to the broader goals of sustainability, resilience, and quality of life in our communities.

In summary, while challenges remain, the prospects for IoT-based solar streetlight systems are bright. Through ongoing research, development, and collaboration, we can realise the full potential of this technology to positively impact human factors and ergonomics in outdoor environments, paving the way for smarter, more sustainable cities and communities.

REFERENCES

Aarthi, M. A. S., & Rajeswari, N. (2022). IoT based automatic street light control and fault detection. *International Journal of Engineering Technology and Management Sciences*, *6*(5), 858–863. https://doi.org/10.46647/ijetms.2022.v06i05.133

Al-badani, M. A., Chong, P. L., & Lim, H. S. (2023a). Enhancing microbial fuel cell performance with carbon powder electrode modifications for low-power sensors modules. *International Journal of Renewable Energy Development*, *13*(1), 80–87. https://doi.org/10.14710/ijred.2024.58977

Al-badani, M. A., Chong, P. L., & Lim, H. S. (2023b). A mini review of the effect of modified carbon paper, carbon felt, and carbon cloth electrodes on the performance of microbial fuel cell. *International Journal of Green Energy*, *21*(1), 170–186. https://doi.org/10.1080/15435075.2023.2194979

Basir, R., Al-Fattah, G., Zainuddin, A. A., Lean, C. P., Kamarudin, S. I., & Aziz, M. A. (2024). Accelerating electric vehicle adoption on Malaysian Islands: Lessons from Japan's Islands of the future initiative. *Lecture Notes in Electrical Engineering 1123*, 109–115. https://doi.org/10.1007/978-981-99-9005-4_14

Beccali, M., Valerio Lo Brano, Bonomolo, M., Cicero, P., Giacomo Corvisieri, Caruso, M., & Gamberale, F. (2017). A Multifunctional public lighting infrastructure, design and experimental test. *Journal of Sustainable Development of Energy, Water and Environment Systems*, *5*(4), 608–625. https://doi.org/10.13044/j.sdewes.d5.0164

Chen, Z., Sivaparthipan, C. B., & Muthu, B. (2022). IoT based smart and intelligent smart city energy optimization. *Sustainable Energy Technologies and Assessments*, *49*, 101724. https://doi.org/10.1016/j.seta.2021.101724

Chong, P. L., Ganesan, S., Ng, P. K., & Kong, F. Y. (2022). A TRIZ-adopted development of a compact experimental board for the teaching and learning of operational amplifier with multiple circuit configurations. *Sustainability*, *14*(21), 14115. https://doi.org/10.3390/su142114115

Chong, P. L., Ganesan, S., Than, Y. Y., & Ravi, P. (2023). Designing an autonomous triggering control system via motion detection for IoT based smart home surveillance CCTV camera. *Malaysian Journal of Science and Advanced Technology* *2*(S1), 80–88. https://doi.org/10.56532/mjsat.v2is1.120

Chong, P. L., Ismail, D., Ng, P. K., Kong, F. Y., Reyasudin, M., & Thirugnanam, S. (2024). A TRIZ approach for designing a smart lighting and control system for classrooms based on counter application with dual PIR sensors. *Sensors*, *24*(4), 1177–1177. https://doi.org/10.3390/s24041177

Chong, P. L., & Kong, F. Y. (2022). System for providing flood and rain alert [Review of system for providing flood and rain alert]. https://iponlineext.myipo.gov.my/SPHI/Extra/IP/Mutual/Browse.aspx?sid=637550536653982775

Chong, P. L., Krishnan, G., Li, C., Kong, N., Ng, N., & Reyasudin, M. (2024). A Raspberry Pi-Powered IoT smart farming system for efficient water irrigation and crop monitoring. *Malaysian Journal of Science and Advanced Technology (Online)*, 149–158. https://doi.org/10.56532/mjsat.v4i2.295

Chong, P. L., & Tan, C. F. (2020). An interactive whiteboard system [Review of an interactive whiteboard system]. https://iponlineext.myipo.gov.my/SPHI/Extra/IP/Mutual/Browse.aspx?sid=637550536653982775

Chong, P. L., Than, Y. Y., Ganesan, S., & Ravi, P. (2022). An overview of iot based smart home surveillance and control system: Challenges and prospects. *Malaysian Journal of Science and Advanced Technology*, *2*, 54–66. https://doi.org/10.56532/mjsat.v2iS1.121

Chong, P. L., Singh, A. K., & Kok, S. L. (2019a). Potential application of Aloe Vera-derived plant-based cell in powering wireless device for remote sensor activation. *PLoS One*, *14*(12): e0227153. https://doi.org/10.1371/journal.pone.0227153

Chong, P. L., Singh, A. K., & Kok, S. L. (2019b). Characterization of Aloe Barbadensis Miller leaves as a potential electrical energy source with optimum experimental setup conditions. *PLoS One*, *14*(6), e0218758. https://doi.org/10.1371/journal.pone.0218758

Chong, P. L., Singh, A. K., & Kyong, F. Y. (2022). Renewable energy from living plants to power IoT sensor for remote sensing. *ADBU Journal of Engineering Technology*, *11*(1). https://journals.dbuniversity.ac.in/ojs/index.php/AJET/article/view/3613

Dwiyaniti, M., Nitisasmita, K. M., & Tohazen. (2019). Energy efficiency on smart street lighting using raspberry pi based on scada and internet of things (IoT). *Journal of Physics: Conference Series*, *1364*, 012034. https://doi.org/10.1088/1742-6596/1364/1/012034

Ekaette Ifiok Archibong, Simeon Ozuomba, & Ekott, E. E. (2020). Internet of things (IoT)-based, solar powered street light system with anti-vandalisation mechanism. *2020 International Conference in Mathematics, Computer Engineering and Computer Science (ICMCECS)*. https://doi.org/10.1109/icmcecs47690.2020.240867

Ganesan, S., Peng, C., Chen, L., Kong, N., Ng, N., & Reyasudin, M. (2024). IoT-enabled Smart Weather Stations: Innovations, Challenges, and Future Directions. *Malaysian Journal of Science and Advanced Technology*, *4*, 180–190. https://doi.org/10.56532/mjsat.v4i2.293

Hossain, J., Algeelani, N. A., Al-Masoodi, A. H. H., & Kadir, A. F. A. (2022). Solar-wind power generation system for street lighting using internet of things. *Indonesian Journal of Electrical Engineering and Computer Science*, *26*(2), 639. https://doi.org/10.11591/ijeecs.v26.i2.pp639-647

Kang, C. Q., Ng, P. K., & Liew, K. W. (2022). A TRIZ-integrated conceptual design process of a smart lawnmower for uneven grassland. *Agronomy*, *12*(11), 2728. https://doi.org/10.3390/agronomy12112728

Lim, S. H., & Ng, P. K. (2021). The design and development of a foldable wheelchair stretcher. *Inventions*, *6*(2), 35. https://doi.org/10.3390/inventions6020035

Masri, M., Liestyowati, D., Andiyan, A., & Husolihah, A. (2024). Smart energy public street lighting system. *IOP Conference Series: Earth and Environmental Science*, *1301*(1), 012008–012008. https://doi.org/10.1088/1755-1315/1301/1/012008

Manoharan, S., Peng, C., None Chen Li, Kong, N., Ng, N., & Reyasudin, M. (2024). IoT-enabled greenhouse systems: Optimizing plant growth and efficiency. *Malaysian Journal of Science and Advanced Technology* *4*(2), 169–179. https://doi.org/10.56532/mjsat.v4i2.294

Ng, P. K., Jee, K. S., & Lim, S. Y. (2016). Development of ergonomics guidelines for improved sitting postures in the classroom among Malaysian university students. *American Journal of Applied Sciences*, *13*(8), 907–912. https://doi.org/10.3844/ajassp.2016.907.912

Ng., P. K., Peng Lean, C., Jian Ai, Y., Yu Jin, N., & Jeyakumar Nathan, R. (2023). Ergonomic work from home recommendations using TRIZ [Review of *ergonomic work from home recommendations using TRIZ*]. In *Human Factors in Engineering Manufacturing Systems, Automation, and Interactions* (pp. 65–82). Taylor & Francis. https://www.taylorfrancis.com/chapters/edit/10.1201/9781003383444-4/ergonomic-work-home-recommendations-using-triz-poh-kiat-ng-peng-lean-chong-jian-ai-yeow-yu-jin-ng-robert-jeyakumar-nathan?context=ubx&refId=f08c3c55-81f5-4d04-a5ad-447a485f1096

Rao, K. S., Peng, C., Ng, P. K., Kong, F. Y., Reyasudin, M., Ismail, D., & Li, N. C. (2024). AI and ML in IR4.0: A short review of applications and challenges. *Malaysian Journal of Science and Advanced Technology (Online)*, 141–148. https://doi.org/10.56532/mjsat.v4i2.291

Saifuzzaman, M., Shetu, S. F., Moon, N. N., Nur, F. N., & Ali, M. H. (2020, July 1). *IoT Based Street Lighting Using Dual Axis Solar Tracker and Effective Traffic Management System Using Deep Learning: Bangladesh Context*. IEEE Xplore. https://doi.org/10.1109/ICCCNT49239.2020.9225590

Sannasy, K., Peng, C., Kong, N., Ng, N., None Chen Li, Reyasudin, M., & Ismail, D. (2024). Transformative Applications of IoT in Diverse Industries: A Mini Review. *Malaysian Journal of Science and Advanced Technology (Online)*, 130–140. https://doi.org/10.56532/mjsat.v4i2.292

Tan, Y. H., Ng, P. K., Saptari, A., & Jee, K. S. (2014). Ergonomics aspects of knob designs: A literature review. *Theoretical Issues in Ergonomics Science*, *16*(1), 86–98. https://doi.org/10.1080/1463922x.2014.880530

Tan, D. W. H., Ng, P. K., Noor, E. E. M., Saptari, A., Hue, C. C., & Ng, Y. J. (2021). Development and usability testing of a finger grip enhancer for the elderly. *Robotics*, *11*(1), 5. https://doi.org/10.3390/robotics11010005

Tian, Y., & Wang, Y. (2020). Street lamp intelligent monitoring system design based on NB-IoT technology. *Journal of Physics: Conference Series*, *1617*, 012003. https://doi.org/10.1088/1742-6596/1617/1/012003

Handayani, T. P., Hulukati, S. A., Jaya, R., Tiandho, Y., & Abdullah, R. (2019). The prototype of solar-powered building lighting IoT. *IOP Conference Series*, *486*(1), 012079–012079. https://doi.org/10.1088/1757-899x/486/1/012079

Yeow, J. A., Ng, P. K., Tan, K. S., Chin, T. S., & Lim, W. Y. (2014). Effects of stress, repetition, fatigue and work environment on human error in manufacturing industries. Journal of Applied Sciences, *14*(24), 3464–3471. https://doi.org/10.3923/jas.2014.3464.3471

Visual Display Technology

5

D. L. Post and H. L. Task

5.1 INTRODUCTION

Many of the design problems encountered by human factors professionals involve electronic visual displays. Although modern human factors training often covers the psychophysical, perceptual, and cognitive aspects of information display, education about display hardware – which is often needed to understand and solve display design problems and work productively with display engineers – is rare. This problem is exacerbated by the fact that most of the literature on display hardware is written for engineers, and that many aspects of the field are evolving quickly. This chapter summarizes the major human factors parameters that are relevant to electronic visual displays, explains the principles of operation for the display technologies of greatest importance today, describes their characteristics insofar as they affect visual performance, and discusses the most common types of display systems that incorporate these technologies.

DOI: 10.1201/9781003675761-5

TABLE 5.1 Common Measures of Contrast

NAME	*DEFINING EQUATIONA*
Contrast ratio b,c	L_{max}/L_{min}
Relative contrast b,c	$(L_{max} - L_{min})/L_{min}$
Luminance contrast c	$(L_{max} - L_{min})/L_{max}$
Luminance modulation or Michaelson contrast	$(L_{max} - L_{min})/(L_{max} + L_{min})$

a Lss_{max} is the greater of the two luminances; L_{min} is the lesser.
b The contrast ratio and relative contrast are undefined when L_{min} is 0.
c For symbology, L_{max} and L_{min} are replaced sometimes by the symbol and background luminances, respectively, in the first three equations, which creates the possibility of contrast ratios < 1 and negative values in the other two cases.

5.2 DISPLAY HUMAN FACTORS PARAMETERS

5.2.1 Photometric Characteristics

One of the most fundamental display characteristics is brightness. Brightness is assessed usually by measuring its psychophysical correlate, luminance, with a photometer. Another basic characteristic is contrast, which is a ratio of on and off pixel luminances. Table 5.1 lists common measures.

Ambient illumination often reduces display contrast by adding reflected light to that coming from the display. Display screens are usually treated to reduce this problem, although most treatments reduce display luminance.

Grayscale is the number of luminances the display can produce at a fixed setting of the brightness and contrast controls, expressed usually as the base-2 logarithm of that number with units called "bits." For example, "8 bits" of grayscale denote 256 levels. The visual system's response to differing luminances is compressive, so technologies having exponentially increasing video-input-to-luminance response functions come closest to providing equally discriminable grayscale levels (although single steps may not actually be discriminable).

For most display technologies, grayscale limits are not inherent and result instead from built-in digital drivers. The exceptions are technologies having pixels that can only be fully on or off. A few technologies – notably CRT – have analog grayscale, which means their luminance can vary continuously and grayscale is limited, if at all, only by their video input signal. Grayscale can be produced in technologies having only on and off pixels by temporal multiplexing, which involves turning pixels on for varying fractions of a frame period, or by dithering, which is a type of spatial multiplexing that involves turning groups of adjacent pixels on and off so the group average has the desired luminance. Usually, dithering is used only as a supplement to temporal multiplexing. Emissive on-off technologies can be dimmed only by using up some of their grayscale (non-emissives can be dimmed by adjusting the illuminant), so they may exhibit contrast artifacts if they are dimmed.

The angular distribution of light emission for some display technologies is approximately Lambertian, which means their luminance is essentially independent of viewing angle; others depart from this ideal in ways that usually make the display harder to read as one moves off-axis. LCD often exhibit additional anomalies off-axis, namely reduced or even reversed contrast and color shifts, although manufacturers are using increasingly sophisticated methods to reduce these problems.

5.2.2 Spatial Characteristics

"Resolution" is used widely, but inconsistently, to describe the amount of spatial detail a display can produce. Display designers often specify resolution in cycles or pixels per unit distance on the display (e.g. cycles/cm), where a cycle is one complete sine wave. Sometimes, the unit distance is implicitly the display's width or height, in which case the specification gives the total number of cycles or pixels that can be produced. Specifications for television sets often state only the number of raster lines that can be displayed. Although conventional television broadcasting provides only 525 lines in the United States (only 483 are visible, though) and 625 lines in the UK and Europe (575 visible), the capacity to display more implies higher resolution within each line. Psychophysicists usually prefer cycles or pixels per unit angle subtended at the viewer's eye (e.g. cycles/degree) because these measures account for the effects of viewing distance and display size on image appearance. Thus, two displays having equal resolution from a display designer's perspective may have different resolutions from a psychophysicist's perspective and vice versa.

Additional confusion is introduced sometimes by failing to distinguish between the resolution of the display and that of the video system driving it or the image it is displaying. For example, using an SXGA (Table 5.2) graphics card does not give a display SXGA resolution, nor does showing a VGA image yield VGA resolution. Another poor practice is confusing resolution with the amount of visual information in an image. For example, a photograph may be described as having higher "resolution" than a drawing of the same scene. CRT produce images by drawing on a screen with a continuously moving electron

TABLE 5.2 Contemporary Digital Resolution Standards

LABEL	*ASPECT RATIO A*	*RESOLUTION (HXV) IN PIXELS*
QVGA	4:3	320 x 240
VGA	4:3	640 x 480
SVGA	4:3	800 x 600
XGA	4:3	1024 x 768
SXGA	5:4	1280 x 1024
UXGA	4:3	1600 x 1200
SDTVB	4:3 or 16:9	704 x 480
HDTVC	16:9	1920 x 1080
	16:9	1280 x 720

A Aspect ratio is the ratio of image width to height.
B SDTV approximates the resolution of the 525-line analog system used currently in the United States. It requires rectangular pixels in one of two possible sizes to achieve the stated aspect ratios. All other entries in Table 5.2 use square pixels.
C HDTV offers two possible resolutions.

beam. Their resolution is mainly a function of their video bandwidth, refresh rate, and electron-beam spot size. These parameters' effects are described most completely by the spatial modulation transfer function, which graphs the response to sinusoidal inputs as a function of their frequency, in a manner analogous to the frequency-response plots one often sees for audio equipment.

Other display technologies consist of a matrix of discrete pixels that are set to the desired luminances by addressing them one or possibly several rows at a time, with all columns loaded simultaneously – methods called "line by line" and "multi-line" addressing, respectively, or, more generally, "matrix addressing." For these technologies, resolution is simply a function of the number of horizontal and vertical pixels. Another addressing method is to connect each pixel to its own driver. This "direct" addressing technique allows each pixel to be driven throughout each frame, but it is practical only for simple applications, requiring only a few dozen pixels or segments (e.g. watches, calculators, and home appliances) because otherwise the number of connections and drivers becomes prohibitive.

Matrix displays come in two types: passive, which have only a pair of electrodes at each pixel, and active, which add a switch (typically, a thin-film transistor, or "TFT," but sometimes a diode) at each pixel so voltage can be maintained throughout the frame period. Passive-matrix displays are cheaper but can apply voltage to the pixels only while their row is selected. Therefore, their luminance and contrast tend to be inferior to otherwise identical active-matrix displays and diminish as the number of rows increases, thus limiting the number of rows that can be provided.

5.2.3 Temporal Characteristics

It is obvious that changing a display image requires redrawing it, but it is less obvious that, for most display technologies, the image fades unless it is refreshed. (The exceptions are bi-stable technologies, which have pixels that remain on or off until they are switched.) In the United States, television refreshes at 30 Hz by alternately drawing all the odd- and then even-numbered lines at 60 Hz. Each set of lines is a "field," both fields constitute a "frame," and this alternating system is called "2:1 interlacing." (In the UK and Europe, the field rate is 50 Hz, yielding a 25-Hz frame rate.) Interlacing reduces bandwidth requirements while avoiding flicker for typical television images but tends to produce jitter in text. Therefore, computer images are usually non-interlaced (this method is also known as "progressive scanning") and refresh at 50 Hz or more. The rate needed to prevent flicker increases as display luminance, contrast, and angular subtense increase. Sometimes, the rate must also be adjusted to prevent beats with fluorescent lighting from becoming visible.

Display luminance is greatest if the pixels emit throughout each frame, but this can cause image smear if the observer tracks a moving object. Therefore, motion rendition is best for display technologies that produce adequate time-averaged luminance while emitting only for small fractions of a frame period. A display having pixels that cannot switch completely during one frame will exhibit contrast losses for moving images, so it is also important that the pixels respond quickly. Most LCD technologies are troublesome in this regard. Speed is even more crucial for technologies that rely on temporal multiplexing to achieve grayscale or use temporal integration to produce color, because in these cases, the pixels must switch more than once during each frame.

5.2.4 Colorimetric Characteristics

The range of colors a display can produce is called its "color gamut" and is determined by its red, green, and blue (RGB) primaries. The closer the primaries come to being monochromatic, the more saturated they become and the larger the gamut becomes; however, no finite set of primaries can reproduce the entire range of colors humans can perceive. Furthermore, increasing a primary's saturation requires reducing its spectral bandwidth, which typically reduces its luminous efficacy; therefore, primary selection often involves a tradeoff between gamut and luminance.

Television broadcasting uses a standard set of primaries that have been chosen for compatibility with available CRT phosphors. Developers of other color display technologies generally try to match or exceed television's gamut. For emissive technologies, one approach uses a white emitter with a patterned RGB color filter in front; another uses RGB emitters, possibly enhanced by a patterned color filter. For non-emissive technologies, the common approach uses a white illuminant and color filters. Luminous efficacy can often be increased by selecting an illuminant having its radiance concentrated in three RGB wavebands and matching the filters' transmittances to those wavebands.

Every color display has a "white point," which is the chromaticity it produces when its RGB channels are driven equally and may or may not be user-adjustable. Television broadcasting and a few other applications use standard white points, defined in terms of Commission Internationale de L'Éclairage (CIE) chromaticity coordinates. Differences in white points are often at least partly responsible for problems in obtaining accurate color transfer when images are viewed on different displays, printed, or scanned.

Most displays produce color by spatial integration: The screen is divided into RGB subpixels, which subtend a visual angle so small that the diffraction patterns they produce on the retina overlap and mix. Varying the subpixel luminances produces what appear to be full-color pixels having any color an RGB mixture can produce.

A similar method, used by most projection displays, is addition. Separate RGB images are produced by three monochrome displays and superimposed internally or at the screen. Addition can yield higher resolution than spatial integration because every pixel can take on any color, but it requires three displays and is therefore bulkier and more expensive.

Color can also be produced by temporal integration – a technique often called "field sequential": the display generates spatially superimposed RGB fields in rapid sequence, within a single frame period, and the visual system integrates them temporally to form a full-color image. Temporal integration has the same resolution advantage as addition, but requires fast displays, sacrifices luminance because each color field is displayed for no more than one-third of each frame, and the images can break up into their RGB components if there is relative motion between the image and the viewer's eye.

Finally, a subtractive process can produce color: white light passes through a stack of three transmissive displays, and each display modulates the R, G, or B component of the light so a full-color image emerges at the front. Subtraction also has a resolution advantage, but requires three displays, careful design to minimize transmission and contrast losses as the light proceeds through three apertures, thin displays to avoid parallax in direct-view applications, and may require special optics to focus the layers simultaneously in head-mounted and projection applications.

5.2.5 Size and Weight

CRT tend to be the heaviest and bulkiest of the display technologies because they are funnel-shaped glass vacuum tubes. Their shape results from the need to provide space for the electron beam to scan the screen. Their weight increases exponentially with screen size because the glass must be thickened to withstand the atmospheric pressure resulting from the internal vacuum. Most other technologies consist only of thin, lightweight layers sandwiched between two flat pieces of glass, so they are more compact and their weight is proportional with screen size. LCD usually require a backlight, though, which can increase their weight and depth substantially.

5.3 EMISSIVE DISPLAY TECHNOLOGIES

5.3.1 Cathode-Ray Tube (CRT)

A CRT produces images by modulating an electron beam as it scans across a phosphor screen. The electrons are produced by heating a cathode, accelerated toward the screen, and focused and scanned by dynamic electromagnetic or electrostatic fields. Dynamic focusing compensates for the changing distance to the screen as the beam scans. Most CRTs scan the beam in a repeating "raster" pattern, consisting of closely spaced horizontal lines; however, oscilloscope CRTs – and some that are used exclusively to show symbols and line graphics – use a "stroke" pattern that moves the beam as if it were a pen. Stroke writing provides increased luminance (because the phosphor can be excited more times per second) and resolution.

Most color CRTs scan three electron beams across a patterned screen containing RGB phosphor dots or stripes, thus producing color by spatial integration. A metal structure ("shadowmask") behind the screen, having one opening for each RGB triplet, helps ensure that each beam strikes only the appropriate phosphor; adjustable magnets also assist. Older, "delta-gun" CRTs arrange the beams in a triangular pattern and include user-adjustable circuitry to ensure a condition called "convergence" that, if violated, produces color fringes that are noticeable especially around white objects on the screen and reduce resolution. Modern, "in-line gun" CRTs arrange the beams in a row and use circuitry and magnets, which are set at the factory and rarely adjusted afterward, to achieve convergence. Color can also be produced by temporal integration, using a white-phosphor CRT in conjunction with a switchable or rotating color filter – this approach is used sometimes in head-mounted systems.

5.3.2 Vacuum Fluorescent Display (VFD)

The VFD is a type of flat CRT that excites a low-voltage phosphor using electrons produced by heating a cathode grill consisting of fine, heated wires. The flow of electrons is controlled by a wire grid that lies between the cathode and a phosphor-coated opaque anode array. The photons then pass through the grid, cathode, and a cover glass. Another arrangement, which deposits the phosphor on the cover glass, is also used sometimes. The anode array can be matrix addressed or, more commonly, arranged in segments. Usually, a blue-green emitting phosphor is used because it has the highest luminous efficacy, but a good range of hues up to red are available.

VFDs are used widely as simple numeric, alphanumeric, and bargraph displays measuring a few square centimeters in products like VCRs, but they have not been practical to >≈160 cm2 for more general-purpose use because this would require adding internal support to control the gap between the cathode and anode, which would interfere with the electron flow. Grayscale and dimming can be produced by varying the anode voltage. Motion rendition is good due to low response time and the emission is Lambertian, so viewing angle is not a problem. Color can be produced by spatial or temporal integration, but multicolor units consisting of differently colored monochrome areas are far more common.

5.3.3 Field-Emission Display (FED)

The FED is another type of flat CRT, in which a phosphor-coated glass screen is excited by electrons produced by an addressable matrix of tiny cathodes. There are basically two types: low-voltage (40–100 V), which require special phosphors and have not yet matched the luminances available from CRT, and high-voltage (5–10 kV), which use conventional phosphors and can match CRT luminances but are vulnerable to damage and failure caused by arcing. Grayscale and dimming can be produced by temporal multiplexing, cathode modulation, or dithering, although multiplexing supplemented by cathode modulation usually yields the best result. The response time is low and the emission is Lambertian. Color is produced by spatial integration, using RGB phosphors, although temporal integration is also feasible.

5.3.4 Image Intensifier Tube (I^2T)

The I^2T is a unique combination of sensor and display in one package. Contemporary third-generation I^2T consist of three primary elements: (1) a photo-cathode screen, which converts photons to electrons; (2) a micro-channel plate (MCP) that accelerates and multiplies the electrons; and (3) a phosphor screen – coated on a fiber-optic or plain glass faceplate – that converts the multiplied electrons to photons. Typically, a power supply is included to convert battery power to the relatively high voltage needed by the MCP in operation, and an image is focused onto the photo-cathode, which then emits electrons in the general direction of the adjacent MCP. The MCP contains millions of tiny tunnels (micro-channels), which amplify the incident electrons in proportion to their number. After exiting the MCP, they are accelerated toward the nearby phosphor screen by an electric field. The phosphor nearest each micro-channel glows with a luminance that is proportional to the number of incident electrons, thus producing a monochrome image that replicates the image focused on the photo-cathode, except that its luminance is greatly amplified. See Section 5.5.4 on NVGs for information on the typical uses of I^2T.

5.3.5 Light-Emitting Diode (LED)

An LED is basically a solid-state material – typically gallium phosphide or gallium arsenide – that has been treated so it consists of two layers that form a P-N junction diode, and passing a current across the layers causes them to emit light. A reflector is placed behind the diode to direct rear-emitted light toward the front, and a lens is often incorporated at the front to further shape the output. The diodes can be arranged in segments to form numeric, alphanumeric, or bargraph displays or laid out as an addressable matrix. A full range of colors, including RGB, is available, so color displays using spatial integration can be created. Grayscale and dimming can be produced by varying the current or temporal multiplexing. The reflectors and lenses are usually designed to produce non-Lambertian emission, to increase on-axis luminance. Large matrices of LED for use as general-purpose desktop color displays have not proven to be practical, mainly because of the difficulty of assembling discrete devices at the required densities (producing different colors on one substrate has proven impossible thus far), but are used widely as message signs. Numeric units – which were once common in wristwatches and calculators until being replaced largely by LCD – are used in clocks and, along with alphanumeric and bargraph units, as instrument readouts.

5.3.6 Organic Light-Emitting Diode (OLED)

An OLED display consists of a thin transparent layer of organic semiconductors or polymers, sealed usually between two glass or plastic substrates bearing orthogonally oriented electrodes on their inner surfaces. The front electrode is transparent and the rear is reflective. Applying a current across the electrodes causes the organic layer to emit visible light. Plastic substrates may make it possible to produce flexible displays, which can be rolled up or folded. Silicon substrates have also been used in experimental, miniature active-matrix OLED designed for head-mounted applications. Currently, this technology is under development, and although it is regarded widely as very promising, there are few commercial products. Grayscale and dimming are produced by varying the electrode current. Viewing angle is good due to Lambertian emission. OLED materials capable of producing RGB light are available, so full color can be produced by spatial integration or, if the substrates are stacked, by addition; however, in the latter case, the stack must be very thin to avoid parallax.

5.3.7 Electroluminescent (EL) Display

The most common type of EL display today uses alternating current to excite a thin-film EL phosphor, yielding the "ACTFEL" structure. The phosphor is sandwiched between thin insulating layers, and this multi-layer structure is deposited typically on a glass substrate bearing transparent electrodes. Reflective, orthogonally oriented electrodes are then applied, and the phosphor is viewed through the glass. Thus, only one substrate is required, and the display can be substantially thinner and lighter than other flat-panel technologies. Ceramic and silicon (supported by ceramic) substrates are also used, although these designs require a cover glass. Miniature devices using active-matrix addressing have been produced for head-mounted use. The angular emission is Lambertian. Grayscale and dimming for EL use temporal multiplexing, enhanced sometimes by dithering. Color is achieved by spatial integration, using either patterned phosphors or a white phosphor plus color filter array.

5.3.8 Plasma Display Panel (PDP)

A PDP consists mainly of a gas that is sealed between two glass substrates bearing orthogonally oriented electrodes on their inner surfaces and, usually, a reflective coating at the rear. Passing a current across the electrodes ionizes the gas, producing a plasma that emits light. Monochrome PDPs use neon, which emits reddish-orange light. In color displays, an ultraviolet-emitting gas is used, and the ultraviolet excites phosphors that are coated on the inner surface of one of the substrates, Thus, a color PDP is similar to a fluorescent light bulb. Either alternating or direct current can be used, but the former is more common today. PDPs are inherently bi-stable, so grayscale and dimming are produced by temporal multiplexing, enhanced sometimes by dithering. Response is very fast, so motion rendition is good. Color is produced by spatial integration. The emission angle is Lambertian for color panels but is isotropic for monochrome panels, causing luminance to increase with viewing angle.

5.4 NON-EMISSIVE DISPLAY TECHNOLOGIES

5.4.1 Illumination Sources

Non-emissive displays require a light source. For some applications, such as LCD wristwatches, ambient illumination reflected off the display may be sufficient, but usually a built-in source is needed. Most direct-view LCDs use one or more tubular fluorescent lamps, mounted in a box containing a reflector or coupled to a light guide. Other approaches that provide more uniform illumination have been tried, though, including EL panels, flat and serpentine-shaped fluorescent lamps, and CRTs. Projection systems require more light than can be provided compactly by a fluorescent source; they use xenon or metal-halide arc lamps. Xenon has a broad spectrum that must be filtered heavily to produce acceptable RGB primaries, thereby reducing overall luminous efficacy. Metal-halide lamps emit mainly in narrow wavebands dictated by the choice of halides, thereby reducing the need for spectral filtering; the fill gas is unavoidably excited too, though, adding at least some broad-spectrum light. Compact, solid-state lasers producing RGB wavelengths have become available recently. Their narrowband spectra are advantageous for many display purposes, and their prices, efficiencies, lifetimes, sizes, weights, and cooling requirements are moving rapidly to a point that may make them attractive for projection uses.

5.4.2 Liquid Crystal Display (LCD)

Liquid crystals (LCs) are elongated molecules that flow like liquids at typical ambient temperatures and tend to align with one another. There are many types of LCs and ways of using them to make displays, but the most common today are nematics, sealed between two pieces of glass that have been treated so the LCs lie parallel with the glass surfaces and undergo a rotation top to bottom, forming a helix – a configuration called "twisted nematic" (TN). In a transmissive TN-LCD, polarizers and transparent electrodes are added to the glass surfaces, and light passing through one polarizer rotates with the twist until it strikes the second polarizer, which either passes or absorbs it, depending on whether the polarizers are aligned with each other or crossed. Applying a voltage across the electrodes rotates the LC so they are perpendicular to the glass, destroying the twist and thereby reversing the second polarizer's effect. Intermediate voltages produce intermediate transmittances and, hence, grayscale. Reflective TN-LCD have a reflective surface at the rear and only a front polarizer, but work basically the same way. The electrodes can be segmented or arranged in a matrix.

Most passive-matrix LCDs today use a twist ranging from 180 to 270° called a "supertwisted nematic" (STN) configuration, which increases the number of rows that can be addressed but reduces viewing angle, response speed, and grayscale. Active-matrix LCDs (AMLCDs) provide the best viewing angle, speed, and grayscale, but are more costly. Other important types of LCDs today are ferroelectrics (FLCs), which are bi-stable and respond to control voltages very quickly but have difficulty producing grayscale, polymer-dispersed (PD) and nematic curvilinear aligned phase (NCAP) LCDs, both of which vary between transparent and light-scattering states and use no polarizers but have trouble producing high contrasts, and polymer stabilized cholesteric texture (PSCT) LCDs, which are bi-stable and therefore do not produce grayscale readily.

5.4.3 Digital Micromirror Display (DMD)

A DMD is an addressable array of tiny mirrors, each of which is mounted on a flexible stalk. Each mirror can be rotated very quickly (e.g. 10 ms) to either of two orientations, thereby deflecting incoming light out to a projection lens or an absorber. Thus, each mirror is a pixel that can be either on or off. Grayscale is produced by temporal multiplexing; dimming is accomplished by modulating the illumination. Color is produced usually by temporal integration using a filter wheel, but triple-DMD systems using addition and dual-DMD hybrids are also available. DMDs do not lend themselves to direct-view applications because of the need for a projection lens; furthermore, it is impractical at the moment to make them bigger than a few square centimeters in area. They are used widely in projectors, though, and head-mounted applications are also possible.

5.5 OPTICAL DISPLAY SYSTEMS

5.5.1 Projectors and Screens

Projection displays use one or more lenses to enlarge and focus an image of an internal display onto a screen. Rear projectors usually incorporate the screen into the housing and therefore tend to be bulky, whereas front projectors use a separate screen and are therefore more compact but require the screen to be distant from the projector. Rear projection screens are usually treated to reduce reflections from room lighting and incorporate lenses to produce a non-Lambertian light distribution that reduces the image viewing angle but provides greater luminance within the intended angle – a feature called "gain." Front projection screens can also have gain.

Some projectors use CRTs to create an image; others, termed "light valves," use non-emissive displays. The least expensive, lightest, and most compact design for a color projector uses a single display and lens system; field-sequential color is most common in self-contained versions, whereas spatial integration is the norm in LCD overlays for overhead projectors (subtractive overlays have also been produced). Greater luminance can be obtained, though, by using three displays to produce separate RGB images that are combined additively; this design is more common in self-contained systems. Most triple-display projectors use dichroic beamsplitters, which reflect one band of wave-lengths while passing the rest, to add the images internally and, in light-valve projectors, to first separate white light into R, G, and B so each component can illuminate the appropriate display. Most CRT projectors, however, use a separate projection lens for each CRT and add the RGB images at the screen.

The most important human factors issues for projectors are resolution, luminance, contrast, optical distortion, viewing angle, size, and weight.

5.5.2 Head-Up Display (HUD)

As the name implies, a HUD is intended to permit the operator of a vehicle (originally an aircraft and, more recently, an automobile) to view a display without looking down toward the instrument panel. It does this by superimposing a display image on the outside world. The original intent was only to assist weapon aiming, but another advantage has been recognized since: the virtual image is normally at or near optical infinity, so the operator does not have to refocus when switching from viewing the world scene to viewing the display.

A HUD consists of three major components: (1) an image source (e.g. a CRT); (2) optics to produce a virtual image of the image source; and (3) an optical combiner, which combines the virtual image with the directly viewed exterior world scene. There are basically two types of HUDs: those that produce a real exit pupil and those that do not. If the HUD uses relay optics, it has a real exit pupil, which means at least one eye must be within the exit pupil to see any part of the display. If the HUD uses simple magnifying optics, it does not have a real exit pupil, and eye position is less critical, although at least one eye must still be positioned within a limited volume of space to see the entire display.

The most important human factors issues for HUD are the combiner ratio (which affects the relative luminances of the symbology and outside world scene), viewing location, exit pupil size or viewing volume, symbology size, type, luminance, and color, stray reflections, and the size and shape of the field of view.

5.5.3 Head- and Helmet-Mounted Display (HMD)

An HMD consists typically of three major components: (1) a miniature display; (2) optics to convey a virtual image of the display to the eye(s); and (3) some (usually adjustable) means of mounting the display and optics to the head. The optics may or may not include a combiner. If the HMD is meant to provide an image that can be superimposed on the external world scene, the optics must include a see-through combiner; if the HMD is for "immersion" only, a combiner is unnecessary and the optics are easier to design. Most HMDs use miniature CRTs or LCDs, which must have very high resolution (in terms of cycles/mm or pixels/mm) to produce a satisfactory image after magnification by the optical system. One exception is the "virtual retinal display," which products an image by scanning a modulated laser beam directly across the retina in much the same way a CRT scans an electron beam across a phosphor screen. Another exception scans an image of an LED array back and forth across the retina.

HMD can be monocular, bi-ocular, or binocular; monochrome, multi- or full-color; see-through or non-see-through; adjustable focus or fixed focus; and real exit-pupil forming or non-pupil forming. The most important human factors issues associated with them are weight, comfort, size, luminance, binocular alignment, resolution, optical distortion, optical distance of the image, size and shape of the field of view, exit pupil size or viewing volume, and the distance the eye can be from the optics ("eye relief").

5.5.4 Night Vision Goggles (NVGs)

NVGs are a special type of HMD. Most NVGs are binocular, with each ocular consisting of three major components: (1) the objective lens, which produces an optical image on the input side of an (2) I2T; and (3) an eyepiece lens to produce a virtual image of the I2T's output. NVGs typically include assemblies that permit them to be adjusted and worn on the head. NVGs containing second-generation I2T are most sensitive to wavelengths from 400 to 750 nm; third-generation devices are most sensitive to 650–900 nm. The most important human factors issues associated with NVGs are ocularity (binocular, bi-ocular, or monocular), output luminance, signal to noise ratio, optical distortion, eye relief, gain, adjustability (fore/aft, interpupillary distance, focus, tilt, up/down), weight, size, and comfort.

5.5.5 Three-Dimensional (3D) or Stereoscopic Display

"3D" is used sometimes to refer to images seen on a conventional display that include depth cues such as perspective, surface shading, and hidden lines. A stereoscopic 3D display, however, provides binocular disparity, so each eye sees a slightly different image that corresponds to what it would see in the real scene. There are several ways to produce binocular disparity. One is to use two independent image sources (e.g. CRT) and an optical system that presents a different image to each eye. Another approach is to use a single display and encode or multiplex the left- and right-eye images so each eye sees only the appropriate one. A common encoding technique, used with projection displays, polarizes the left- and right-eye images oppositely and superimposes them on the screen; the observers wear oppositely polarized lenses, so each eye sees only the appropriate image. A common multiplexing technique, used with direct-view displays, places electronic shutters in front of the eyes and alternates the left- and right-eye images on the display in synchrony with the opening and closing of the shutters.

The most important human factors issues associated with stereoscopic displays are the accuracy of the binocular disparities and hence depth cues, visibility of the images to the wrong eyes ("crosstalk"), frame rate, luminance, color, binocular alignment, and the effects of mismatch and lack of normal covariance between the images' optical distance (which is usually fixed) and the varying vergence cues and requirements produced by the binocular disparities.

ACKNOWLEDGMENTS

The authors thank Howard L. Funk (H.L. Funk Consulting), Bernard Lechner (consultant), Jeffrey Ohstrom (Thomson Components and Tubes Corp.), and Lawrence Tannas Jr (Tannas Electronics) for advising us on several important matters, Reginald Daniels and Gurdial Saini (Air Force Research Laboratory) for reviewing the FED and OLED sections, and especially Alan Sobel (Editor of *Journal of the Society for Information Display*) for reviewing the manuscript.

6 Visual Display Height

R. Burgess-Limerick

6.1 PREFERRED GAZE ANGLE AND VERGENCE DEMANDS

Many workstations require fixation on relatively close visual targets, such as computer displays. These visual targets should be located at least 15° below horizontal eye level.

The argument for lower visual targets commences with the observation that subjective preference is for visual targets to be located such that the eyes are rotated downwards relative to the head. Kroemer and Hill (1986), for example, measured the average preferred gaze angle as 35° below a line joining the external auditory meatus and the outer canthus (the ear-eye line) for visual targets at 1 m, and 44° below the ear-eye line for targets at 0.5 m, while Mon-Williams et al. (1999) found the preferred gaze angle of 12 participants to range from 19° to 36° below the ear-eye line (mean, 27°) for a target at 0.65 m.

A mechanical mechanism for this phenomenon has been proposed based on the knowledge that an observer must converge to maintain single vision of near visual targets. This ocular vergence is produced by activation of the medial recti muscles of the eye. However, the extraocular muscles that raise the eyes (the superior recti and inferior obliques) also create a horizontal divergent

DOI: 10.1201/9781003675761-6

force. Raising the eyes thus increases the activation required of the medial recti which causes visual discomfort. This simple mechanical model explains why observers prefer to look downwards to view near targets, and why the preferred vertical gaze angle gets progressively lower for closer objects. Measurements of open loop heterophoria (an indirect measure of vergence effort) as a function of vertical gaze angle are consistent with these conclusions (Mon-Williams et al. 1998, 1999).

6.2 INFLUENCE OF TARGET LOCATION ON GAZE ANGLE AND POSTURE

In a normal erect posture, the ear-eye line is typically about 15° above the horizontal. Consequently, for a seated observer to fixate a visual target placed at horizontal eye height, either the preferred gaze angle must be compromised (leading to increased vergence effort) or the head rotated posteriorly by some combination of atlanto-occipital or cervical extension. It has been demonstrated experimentally that changing the height of visual targets results in changes in both gaze angle and head orientation, and the changes in head orientation associated with changes in vertical target location are achieved predominantly by changes in atlanto-occipital posture. Gaze angles higher than preferred were adopted when the visual target was higher than 15° below horizontal eye level (Figure 6.1).

6.3 ATLANTO-OCCIPITAL AND CERVICAL BIOMECHANICS

Interpretation of the consequences of the observed postural responses requires consideration of the biomechanics of the head and neck. The head and neck system comprises a rigid head located above a relatively flexible cervical spine. Flexion and extension are possible at the atlanto-occipital and cervical joints. The ligaments and joint capsules are relatively elastic, especially within the mid-range, and a large range of movement is possible without significant contribution from passive tissues. The center of mass of the head, and the head and neck combined, is anterior to the atlanto-occipital and cervical joints. Consequently, extensor torques about the atlanto-occipital and cervical

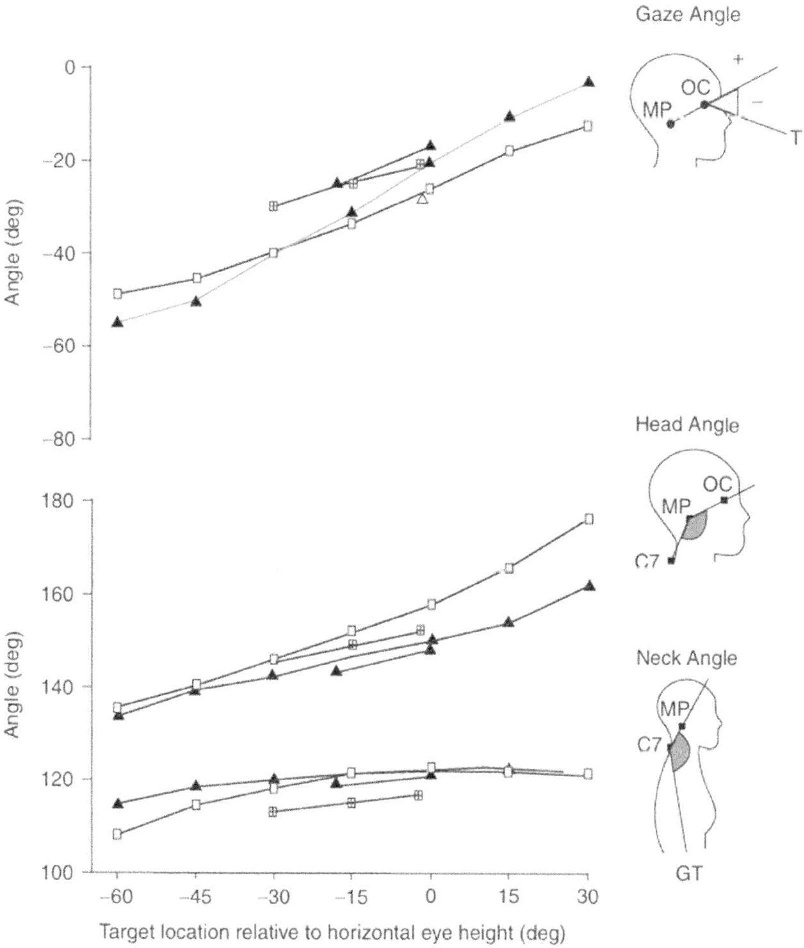

FIGURE 6.1 Gaze angle relative to the ear-eye line, head angle, and neck angle as a function of target location relative to horizontal eye height. Average data for each of four experiments (Burgess-Limerick et al. 1998, 1999, 2000; Mon-Williams et al. 1999).

joints are required to maintain static equilibrium when the trunk is vertical. A large number of muscles with diverse sizes, morphology, and attachments are capable of contributing to these torques. The suboccipital muscles, which have origin on C1 and C2 and insert on the occipital bone, are capable of providing extensor torque about the atlanto-occipital joint only; others (such as semispinalis capitis) provide extensor torque about the cervical as well as the atlanto-occipital joints; while others provide extensor torque about cervical vertebrae only.

Increased flexion at the atlanto-occipital joint increases the horizontal distance of the center of the mass of the head from its axis of rotation (level with the mastoid process). Similarly, when the trunk is approximately vertical, an increase in flexion of the cervical spine increases the horizontal distance of the center of the mass of the head and neck combined from the axes of rotation in the vertebral column (and all else remaining the same, the horizontal distance of the head from its axis of rotation). Hence, with the trunk approximately vertical, both atlanto-occipital and cervical flexion increase the torque required of the extensor musculature to maintain static equilibrium. The conventional view, based on the analysis above, is that an erect head and neck posture that reduces the flexor moment of the head is to be preferred. Whilst such a simplified model of the situation is intuitively attractive, the cervical spine is particularly complex. The result is that there is no definitive answer to the question of what is an optimal posture (or range of postures) of the head and neck.

A recommendation to avoid static postures involving neck flexion greater than 30° from upright can be justified by an experiment conducted by Chaffin (1973), in which the time taken to reach significant muscle fatigue decreased from 5 to 2 h when neck flexion increased from 30 to 60°. The consequences of flexion of a lesser degree are less certain, however. Electromyographic evaluations of different neck postures yield varied results. Schuldt et al. (1986), for example, demonstrated elevated electromyographical activity in the posterior neck musculature during maximal flexion but no significant differences for intermediate postures, while Harms-Ringdahl et al. (1986) did not detect electromyographic differences between neutral and extreme flexion. Kumar (1994) reported decreased trapezius EMG and subjective discomfort associated with increased neck inclination and flexor moment. Turville et al. (1998) also reported a decrease in mean trapezius activity when participants worked at a computer monitor lowered by 25°, but found an increase in mean cervical erector spinae activity. These apparent anomalies arise because the external flexor moment is only one factor contributing to the demands placed on the neck musculature. The head and neck is an inherently unstable system – especially in the upright position (Winters and Peles 1990). Consequently, the neck muscles must do more than just balance the external forces acting on the system. For the system to be stable, additional co-contraction is required to increase the stiffness of the cervical spine and prevent buckling. The consequence is that significant muscular activity is probably required to stiffen the cervical spine, even if the head and neck are positioned to minimize the flexor torque imposed by gravitational acceleration. Indeed, the necessity for muscle activity to stabilize the cervical spine is likely to be greater when it is relatively extended (Winters and Peles 1990).

Another factor that must be included is the recognition that the tension-generating capability of a muscle is highly dependent on its length. In general, changes in posture at the atlanto-occipital and cervical joints will alter both the moment arm and the average fiber length of the muscles actively providing both the required extensor torque and stiffness. While accurate measurements of moment arm and fiber length changes are unavailable, it is clear that the muscle fibers that produce extensor torque will be shortened to some extent by increased extension of the head and neck. The suboccipital muscles in particular are relatively short, and even a small change in average fiber length caused by extension of the atlanto-occipital joint is likely to cause a significant decrement in their tension-generating capabilities. Yet it is precisely these muscles which appear to be primarily responsible for vertical movements about axes high in the cervical spine (Winters and Peles 1990). The best available estimates suggest that extension of the atlanto-occipital joint beyond neutral rapidly leads to a decrement in the force-generating capability of the small suboccipital muscles (Burgess-Limerick et al. 2000). This is also true of muscles which cross both cervical and atlanto-occipital joints (such as the semispinalis capitis), unless the cervical spine is in a markedly flexed posture.

6.4 CONCLUSION

In a normal erect posture, the ear-eye line is typically 15° above horizontal (Jampel and Shi 1992), and this provides the best available definition of the neutral posture of the atlanto-occipital joint. When viewing visual targets from a seated position, the head is, on average, held in this erect posture when the visual target is 15° below horizontal eye height. In addition to leading to gaze angles higher than preferred, locating visual displays higher than 15° below horizontal causes extension of the atlanto-occipital joint from neutral. Even a small amount of extension of the atlanto-occipital joint is likely to cause a decrease in the tension generating capabilities of both the suboccipital muscles and the cervical muscles which insert on the occiput (although the gradient of the relationship for these latter muscle groups will depend on the cervical posture). These conclusions support a design recommendation that visual targets should be located lower than 15° below horizontal eye height.

REFERENCES

Burgess-Limerick, R., Mon-Williams, M., & Coppard, V. (2000). Visual display height. *Human Factors*, 42, 140–150.

Burgess-Limerick, R., Plooy, A., & Ankrum, D. (1998). The effect of imposed and self-selected computer monitor height on posture and gaze angle. *Clinical Biomechanics*, 13, 584–592.

Burgess-Limerick, R., Plooy, A., Fraser, K., & Ankrum, D. R. (1999). The influence of computer monitor height on head and neck posture. *International Journal of Industrial Ergonomics*, 23, 171–179.

Chaffin, D. B. (1973). Localized muscle fatigue: Definition and measurement. *Journal of Occupational Medicine*, 15, 346–354.

Harms-Ringdahl, K., Ekholm, J., Schuldt, K., Nemeth, G., & Arborelius, U. P. (1986). Load moments and myoelectric activity when the cervical spine is held in full flexion and extension. *Ergonomics*, 29, 1539–1552.

Jampel, R. S., & Shi, D. X. (1992). The primary position of the eyes, the resetting saccade, and the transverse visual head plane. *Investigative Ophthalmology and Visual Science*, 33, 2501–2510.

Kroemer, K. H. E., & Hill, S. G. (1986). Preferred line of sight angle. *Ergonomics*, 29, 1129–1134.

Kumar, S. (1994). A computer desk for bifocal lens wearers, with special emphasis on selected telecommunication tasks. *Ergonomics*, 37, 1669–1678.

Mon-Williams, M., Burgess-Limerick, R., Plooy, A., & Wann, J. (1999). Vertical gaze direction and postural adjustment: An extension of the Heuer model. *Journal of Experimental Psychology: Applied*, 5, 35–53.

Mon-Williams, M., Plooy, A., Burgess-Limerick, R., & Wann, J. (1998). Gaze angle: A possible mechanism of visual stress in virtual reality headsets. *Ergonomics*, 41, 280–285.

Schuldt, K., Ekholm, J., Harms-Ringdahl, K., Nemeth, G., & Arborelius, U. P. (1986). Effects of changes in sitting work posture on neck and shoulder muscle activity. *Ergonomics*, 29, 1525–1537.

Turville, K. L., Psihogios, J. P., Ulmer, T. R., & Mirka, G. A. (1998). The effects of video display terminal height on the operator: A comparison of the 15° and 40° recommendations. *Applied Ergonomics*, 29, 239–246.

Winters, J. M., & Peles, J. D. (1990). Neck muscle activity and 3-D head kinematics during quasi-static and dynamic tracking movements. In J. M. Winters & S. L.-Y. Woo (Eds.), *Multiple muscle systems: Biomechanics and movement organisation* (pp. 461–480). Springer Verlag.

FURTHER READING

Abdel-Aziem, A. A., Abdel-Ghafar, M. A. F., Ali, O. I., & Abdelraouf, O. R. (2022). Effects of smartphone screen viewing duration and body position on head and neck posture in elementary school children. *Journal of Back and Musculoskeletal Rehabilitation*, 35(1), 185–193.

Cruz, T. L., Pérez, S. M., & Chiappe, M. E. (2021). Fast tuning of posture control by visual feedback underlies gaze stabilization in walking Drosophila. *Current Biology*, 31(20), 4596–4607.

Mrugalska, B., & Karwowski, W. (Eds.). (2023). *Anthropometry: Human body measurements and how to use them*. Taylor & Francis.

Raffi, M., Persiani, M., & Piras, A. (2022). The role of optic flow and gaze direction on postural control. *Emerging Science Journal*, 6(6), 1541–1551.

Stincel, O. R., Oravitan, M., Pantea, C., Almajan-Guta, B., Mirica, N., Boncu, A., & Avram, C. (2023). Assessment of forward head posture and ergonomics in young IT professionals–reasons to worry?. *La Medicina del Lavoro*, 114(1), e2023006.

Visual Displays in Human-Computer Interaction

7

H. Luczak, L. Schmidt, O. Oehme and M. Rötting

7.1 INTRODUCTION

This chapter gives a general overview of past, current, and future visual displays, their technical basis, and the resulting characteristics relevant to human-computer interaction (HCI). In this context, computers may be understood as stationary desktop systems, as well as mobile devices like notebooks, personal digital assistants (PDAs), and other pocket or handheld devices.

Together with the general development of HCI, the technology of visual displays has quickly evolved over the last decades (Luczak et al. 2003). The majority of visual displays in HCI is based on the variation of light (e.g. indicator lights signaling power on or the main visual display), but there are still some visual indicators on common computers that rely on the displacement of a physical entity. Examples are the write-protect tab on a 3.5" floppy disk, the eject button on the floppy disk drive, or latches on notebook computers.

DOI: 10.1201/9781003675761-7

Displays based on the displacement of a physical entity might have the advantage that they can be perceived by touch and therefore do not rely solely on the human visual system to convey their message. However, the amount of information that can be displayed is restricted. Therefore, all displays discussed below are based on visual light, i.e. electro-magnetic radiation with a wavelength between about 380 and 780 nm.

Some newer developments in HCI (e.g. virtual reality and augmented reality) require visual displays with characteristics very different from today's found on standard PCs. On the other hand, technical devices with formerly restricted interaction possibilities are becoming more "intelligent." Hence, user interaction is more diverse, and one way of accomplishing this is by implementing visual interfaces with greater spatial resolution. Many of these devices show a similar evolution of the visual display as the computer did: from simple lamps to graphical displays.

7.2 DEVELOPMENTS IN DISPLAY TECHNOLOGY

7.2.1 Lamps

With the invention of the light bulb, the first digital possibility to change electric power into a defined artificial light was created. Using this technology, it was possible to display two different stages with lamps, e.g. a conference room as "occupied" or as "free." Also, the first freely programmable computer, Konrad Zuse's Z3, used lamps as output to display four decimal digits with decimal point.

Lamps can be classified into incandescent and gas discharge lamps, based on the physical principle of light generation. In the incandescent lamp, a current is passed through a thin wire filament, which causes the wire to glow to white heat due to resistance. To prevent the filament from burning, it is sealed in an evacuated glass bulb. The spectrum of the emitted light is a function of the temperature of the filament, from red (1500 K) over yellow and white to blue (6500 K). Incandescent lambs have a low efficiency. Only about 5% of the energy is emitted as light, and the rest is emitted as heat.

In gas discharge lamps, the atoms or molecules of the gas inside a glass, quartz, or translucent ceramic tube are ionized by an electric current passing through the gas. Photons are emitted when electrons of the gas molecules, temporarily displaced from their respective energy levels, return to their old energy level. Most photons emitted are within the ultraviolet spectrum, so an additional transformation takes place in a coating inside the tube, transforming ultraviolet into visible light. The color of gas discharge lamps depends mainly on the mixture of gasses and the coating of the tube.

Lamps used as displays are very often colored to make them more conspicuous or as a means of coding.

7.2.2 Light-Emitting Diodes

In many applications, light-emitting diodes (LEDs) replaced lamps because of their longer mean time between failures, smaller size, and less energy consumption. In the 1960s and 1970s, the LEDs matured and became the standard display, first for the electronic calculator and shortly after for the watch. In both markets, LEDs were eventually replaced by LCDs. Like lamps, their role in HCI is currently restricted to a simple status indicator for power or hard disk access.

However, recent technical breakthroughs, among them the availability of LEDs in just about any color, led to the development of large public displays based on LEDs, and work is going on in developing full-color miniature LED matrices for head-mounted and head-up displays.

7.2.3 Cathode Ray Tubes

The next generation of computers, with line-oriented interfaces, utilized monochrome cathode ray tubes (CRTs) to display alphanumerical characters (Chadha 1995). A CRT systematically consists of a glass bulb, a cathode, an electron gun, deflection coils, a mask, and a phosphor coating. The electron beam is directed line-by-line across the screen and stimulates the phosphor coating to lighten up for a short moment.

With the advent of full-screen interfaces, CRTs capable of displaying grayscales, colors, and graphics became common. Technically the grayscales of CRTs were realized with a modulated intensity of the light which the voltage applied to the cathode. The more voltage was applied, the brighter the light shone. Color representation is reached by the kind of phosphor that is used and the electron beam. The phosphorus layer is made of a "triad" in the form of an array, with every triad consisting of a dot of each of the colors red, green, and blue. Therefore, color monitors have three electron guns. The electron beam of each of them only falls on one colored dot in each triad (Bosman 1989). Three dots, each of one triad, are one pixel. With the help of different light intensities, almost every mixture of colors is possible. With today's direct manipulation operating systems, displays capable of generating graphics with a high resolution and millions of colors are the standard for personal computers.

7.2.4 Liquid Crystal Displays

Together with the development of laptop computers, small and light liquid crystal displays (LCD) also became an alternative to CRT for desktop PCs. Results from experiments have shown that people expressed better readability and less mental strain using an LCD instead of a CRT. Other advantages of LCDs are their minimal power consumption, their operation at low voltages, their longevity in normal environments, and the fact that displays may either be viewed directly in transmission or reflection, or may be projected on to large screens (Bosman 1989).

LCDs consist of two glass plates with microscopic lines or grooves on the inner surface, and a liquid crystal layer located between them. Liquid crystal materials do not emit light of their own, so some means of external or back illumination have to be provided. The physical principle is based on the anisotropic material qualities of liquid crystals. When substances are in an odd state – partly liquid and partly solid – their molecules tend to point the same way, like the molecules in a solid, but can also move around to different positions, like the molecules in a liquid. This means that liquid crystals are neither a solid nor a liquid but they are closer to a liquid state. In an electric field, liquid crystals change their alignment and therefore their translucence. If no voltage is set up, the light can pass through, and the pixels appear bright. In areas with voltage set up, the pixels become dark.

Traditional pixel-by-pixel addressing results in poor contrast image. Each pixel is addressed by a passive matrix, which consists of horizontal and vertical electron stripes. Thus, each point of intersection represents a subpixel that can be addressed. The electric field, however, expands over the pixel to be addressed to the entire horizontal and vertical electrode, respectively. This results in disturbing stripes, so-called ghosting. Other disadvantages are slow response times and a prelightening of the pixels.

Colors can be produced by RGB filters placed in front of the pixels. A disadvantage is the relatively small translucence caused by the color filter and the polarizers.

7.2.5 Thin Film Transistor Displays

In recent years, apart from larger displays, there has been a general shift from stationary desktop-based general-purpose computers to mobile and more task-specific information appliances. Notebooks, pocket computers, personal digital assistants (PDA), and mobile phones have become more and more the focus of research as their commercial dominance has grown. People take their notebooks, phones, and PDAs with them everywhere, using them in various environments and situations to perform different tasks. Therefore, new requirements are, for example, the changing and unfavorable ambient light conditions, different viewing angles, and faster switching capabilities to display moving objects and videos.

By depositing an active matrix of thin film transistors (TFT) on the LCD, one per pixel, the problems of traditional (dual scan twisted nematic [DSTN]) LCDs with passive activation can be overcome. By turning on these transistors, the charge can be maintained, and the display contrast remains stable. In order to display a full range of colors, each element on an LCD is assigned to a primary color by a special filter which is deposited on the face plate of the LCD. To guarantee high display contrast quality, brightness, and high quality of color representation, many transistors are needed. For a resolution of 1024 x 768 pixels, 2.36 million transistors are required – one per primary color and subpixel.

TFTs also improve the width of the viewing angle and reduce the switching latency. Today, this display technology is the widest spread variant of LCDs.

7.2.6 Plasma Display Panels

Especially for video conferencing or group presentations, which require the lights to remain on, plasma display panels (PDPs) are in use. Unlike front view projection screens, the lights do not have to be turned off to see the image clearly and easily.

Light is produced by an electrical discharge in a gas. Two glass plates are laid with their parallel, thin conducting paths right-angled, one upon the other. The gap between the plates is evacuated and filled with a gas mixture. If sufficient high voltage is applied to the cross point of two orthogonal conducting paths, the gas ionizes and begins to shine (like a mass of little gas discharge lamps). However, color representation is not easily obtained. The inside of one glass plate is coated with a phosphorus layer, which is – according to the fundamental colors – composed of three different kinds of phosphorus.

The advantage of this technology is the flat construction of large screens which perform extraordinarily well under most ambient light conditions. Very bright light does not wash out the image on the screen. Another characteristic of a plasma panel is the extreme viewing angles both vertically and horizontally.

With 160° of viewing angle, people sitting to the side of the plasma screen will still be able to see the image without losing any information.

Its disadvantage is, firstly, its high consumption of electricity. This technology is unsuitable for battery-operated and portable computers Further, there is the necessity of high voltage to ignite the plasma. Moreover, the phosphorus layers are exhausted after a period of use. Finally, the pixels are oversized, requiring the user to keep a distance from the display.

7.2.7 Virtual Retinal Displays

In the field of virtual and augmented reality (VR, AR), head-mounted displays (HMDs) are often used to provide virtual and real world information. VR can be defined by the total immersion of the viewer into an artificial "virtual" world, with the user no longer having a view of the real world. This means that the left and the right eye are provided with different image information. Thus, a three-dimensional view is possible. A commonly used HMD has two small displays (LCD or CRT) producing the image and offering it via special optics to the eyes. In contrast to VR, AR involves the enrichment of the real world with virtual information (Azuma 1997). To be able to see the real world as well as the virtual display information, HMDs with integrated LCDs such as for VR but with half-silvered mirrors, are used to realize the superimposition of the real world with virtual information. However, such HMDs do have disadvantages, e.g. the bad see-through features (rather like looking through sunglasses), their large size and weight, or their limited field of view (FOV) of about 30°. Actions taking place outside the FOV cannot be perceived by the user due to the construction of the HMD, resulting in a limited peripheral view of the real world.

To overcome these disadvantages, new display generations, such as virtual retinal display (VRD) technology (also called retinal scanning display [RSD]) have been developed. The VRD addresses the retina directly with a single stream of pixels. The line diversion, horizontal and vertical, is reached by the use of rotary or oscillating mirrors. Light output from a laser is modulated according to the input signal, and the red, green, and blue light is combined. This combined white light is then raster-scanned onto the retina to create an image. Thus, a very clear and sharp projection of different kinds of information can be guaranteed (Microvision 2001). In addition, an additional projection surface inside the glasses becomes unnecessary. Due to the higher light intensity of a laser, the half-silvered mirrors commonly used for HMDs can be optimized to a maximum translucence. Thus, the see-through qualities of the glasses can be improved. Furthermore, the display's maximum resolution is no longer determined by the tolerances of manufacturing of the display pixels but by the control logic and deflection mirror's quality. However, current VRDs are still prototypes, and the first products are only just becoming available. For this reason, it will take some time before such devices can be used as a real alternative to LC-based HMDs.

7.3 FROM OUTPUT-ONLY TO DIRECT INTERACTION DEVICES

During the last few years, far-reaching developments concerning new display types, as well as new interaction concepts, have been observed (Luczak et al. 2002). The first visual displays in HCI were, rather like standard computer monitors today, only pure representation or output devices with no possibilities for interaction. The interaction concepts of Windows-based software has become a standard for HCI and is, to a large extent, not dependent on display technology.

However, with the development of more task-specific displays, Windows-based graphical user interfaces are not suitable for various application cases. In particular, for HMDs in VR and AR, new interaction concepts are still required. Several years ago, together with the emergence of smaller and mobile devices such as PDAs and handheld computers, touch-screen displays became the norm, even for stationary use. In contrast to standard windows pointing devices like a computer mouse, this allows more task-specific and direct HCI. Pointing is no longer in another plane than the visualized interactions on the display. Nowadays, tablet PCs enable input and output of handwriting and pen-based interactions using touch-screen displays in a more natural way. However, the pen as a pointing and input device also requires new interaction concepts, e.g. for left- and right-handed users, minimizing hand movements, or detecting gestures. New applications are often technology-driven, but human-oriented consequences of different ways of interaction have to be taken into account.

The approach to directly stimulate the retinal receptors by a VRD gives way to the assumption that other receptors and even brain cells can be directly triggered in the future. Thus, it is imaginable that conventional displays will cease to exist and will be replaced by chips which can directly control the nerve centers of the brain and create a virtual image.

REFERENCES

Azuma, R. (1997). A survey of augmented reality. *Presence*, 6, 355.

Bosman, D. (Ed.). (1989). *Display engineering: Conditioning, technologies, applications*. Amsterdam: Elsevier Science Publishers.

Chadha, S. S. (1995). *An overview of electronic displays.* London: Applied Vision Association.

Luczak, H., Cakir, A. E., & Cakir, G. (Eds.). (2002). WWDU 2002 – work with display units – world wide work. In *Proceedings of the 6th International Scientific Conference, Berchtesgaden, May 22–25.* Berlin: ERGONOMIC Institut für Arbeits- und Sozialforschung.

Luczak, H., Rötting, M., & Oehme, O. (2003). Visual displays. In J. A. Jacko & A. Sears (Eds.), *The human–computer interaction handbook — Fundamentals, evolving technologies and emerging applications* (p. 187). Mahwah, NJ: Lawrence Erlbaum Associates.

Microvision. (2001). Homepage Microvision Inc. Retrieved August 3, 2001, from http://www.mvis.com/

FURTHER READING

Chai, C., Shi, J., Wu, C., Zhou, Y., Zhang, W., & Liao, J. (2022). When to use vibrotactile displays? A meta-analysis for the role of vibrotactile displays in human–computer interaction. *Applied Ergonomics*, 103, 103802.

Erickson, A., Kim, K., Lambert, A., Bruder, G., Browne, M. P., & Welch, G. F. (2021). An extended analysis on the benefits of dark mode user interfaces in optical see-through head-mounted displays. *ACM Transactions on Applied Perception (TAP)*, 18(3), 1–22.

Han, L., Zhang, H., Xiang, Z., Shang, J., Anjani, S., Song, Y., & Vink, P. (2021). Desktop lighting for comfortable use of a computer screen. *Work*, 68(s1), S209–S221.

Kim, Y. J., Kumaran, R., Sayyad, E., Milner, A., Bullock, T., Giesbrecht, B., & Höllerer, T. (2022). Investigating search among physical and virtual objects under different lighting conditions. *IEEE Transactions on Visualization and Computer Graphics*, 28(11), 3788–3798.

Zhu, Y., Geng, Y., Huang, R., Zhang, X., Wang, L., & Liu, W. (2024). Driving towards the future: Exploring human-centered design and experiment of glazing projection display systems for autonomous vehicles. *International Journal of Human–Computer Interaction*, 40(15), 4087–4102.

Noise *Definitions*

8

Z. Engel, D. Augustynska,
J. Koton and A. Kaczmarska

8.1 BASIC DEFINITIONS

All disturbing, annoying, fatiguing, and hazardous sounds influencing the hearing organ and other senses of a human body are considered noise.

From the point of view of physics, all sounds are mechanical vibrations of an elastic medium (gas, liquid, or solid). These vibrations are considered an oscillating movement of medium particles around the state of equilibrium, causing a change of medium pressure versus static (atmospheric) pressure. Out of this change of pressure (or, the disturbance of equilibrium), which in a form of successive local condensations and thinnings of medium particles is transferred into the space surrounding the vibration source, an acoustic wave is formed.

The difference between the pressure instantaneous value at the acoustic wave passing pressure value is called acoustic pressure, (expressed in Pa). Considering the wide range of an acoustic pressure from 2×10^{-5} to 2×10^{2} Pa, the logarithmic scale is used, and as a consequence the notion of an acoustic pressure level, L, expressed in dB is applied, according to the equation:

DOI: 10.1201/9781003675761-8

$$L = 10\lg\frac{p^2}{p_0^2} \tag{1}$$

where p is root mean square pressure value (Pa), p_0 is threshold of rms pressure value, called the reference sound pressure value, equal to 20 mP.

This is a nominal value equivalent of an acoustic pressure of the single tone (sinusoidal vibration) of 1000 Hz frequency at which the hearing impression forms; this means that the so-called threshold of audibility for that frequency occurs.

Ranges of an acoustic pressure occurring in an environment are shown in Figure 8.1.

All values characterizing the noisiness of work environment, which will be discussed in further sections here – maximal sound A level, peak sound C level, A-weighted sound level, the level of personnel exposure to a noise related either to an 8-h work period or to a week work period – are quantities based on the acoustic pressure level.

The acoustic pressure levels corrected according to the frequency-weighting characteristics A and C sound level meters are called the sound A level and the sound C level respectively.

The equivalent A-weighted sound level, $L_{Aeq,T}$ (the value used either for the description of the noise varying in time or the variable exposure time) is defined as a mean value of the sound A level varying in time being an equivalent to the reaction of a hearing organ exposed to the influence of a constant level noise in a weighted period of time. It is expressed in dB and given by the formula:

$$L_{Aeq,T_e} = 10\lg\left[\frac{1}{T_e}\int_0^{T_e}\left(\frac{p_A(t)}{p_0}\right)^2 \mathrm{d}t\right] \tag{2}$$

where T_e is time of exposure (s), and p_A is instantaneous value of an acoustic pressure, corrected according to the frequency-weighting characteristics $\mathrm{A}(\mathrm{Pa})$.

The A-weighted sound levels, in dB, related to a normal work period (either 8 h or a week) are referred to as sound exposure level related to the 8-h work period $L_{\mathrm{EX,8h}}$ or related to the week work period $-L_{\mathrm{EX,w}}$, and are given by formulae:

$$L_{\mathrm{EX,8h}} = L_{Aeq,T_e} 10\lg\frac{T_e}{T_0} \tag{3}$$

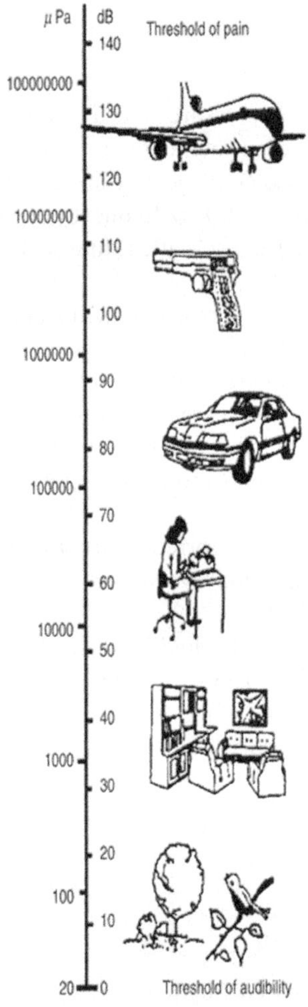

FIGURE 8.1 The maximal sound A level refers to the maximal effective level of sound A occurring during the observation, while the peak sound C level refers to the maximal instantaneous level of sound C occurring during the observation.

$$L_{\text{EX,8h}} = 10\lg\left[\frac{1}{5}\sum_{i=1}^{n}10^{0.1\left(L_{\text{EX,8h}}\right)}\right] \tag{4}$$

where L_{Aeq,T_e} is equivalent A-weighted sound level related to the sound exposure time T_e (dB), T_0 is reference time $= 8\text{h} = 28800\text{s}$, i is consecutive workday in a specified week, and n is number of workdays in a specified week (can be different than 5).

Instead of the exposure time level related to the workday or week, another expression, known as the daily or weekly noise exposure, is used, $L_{A,T_e}\left(\mathrm{Pa}^2\mathrm{s}\right)$. The interrelation between the noise exposure and the noise exposure level related to an 8-h work period is as follows (2-4):

$$L_{A,T_e} - 1.15 \cdot 10^{-5} \cdot 10^{0.1L_{\mathrm{EX,8h}}} \tag{5}$$

Knowledge of several additional acoustic phenomena is needed for the effective application of technical means of noise control. They are:

- Velocity of an acoustic wave propagation (speed of sound), c; this means the velocity of the propagation of disturbance of the medium equilibrium defined as the ratio of the distance traveled by the disturbance in an elementary time interval to the value of that interval. For example, this velocity equals 340 $\mathrm{m/s}$ with air temperature of $20^\circ\mathrm{C}$ and at normal atmospheric pressure.
- Acoustic vibration period, T, is the shortest time interval after which the same state of an observed phenomenon (vibrations or disturbances) repeats itself.
- Acoustic vibration phase, f, is the value of a vibrating particle deviation – in the given point and at the given moment – from the mean value of the particle location.
- Acoustic vibration frequency (sound frequency), f, is the number of vibration periods in the time unit.
- Acoustic wavelength, l, is the distance between two successive points, measured in the direction of the disturbance propagation, in which the vibrations are in the same phase (or the distance traveled by the wave front during one full period).

The acoustic wavelength is given by the expression: $l = c/f$, in m, where c is the sound velocity in $\mathrm{m/s}$, and f is the frequency (Hz). For the audio frequency range, $f = 16 - 16000$ Hz, and the wavelength, $l = 21 - 0.021$ m.

Simplified, one can say that the noise is the sum of a large number of sinusoidal vibrations. Factoring the complex vibrations into a sum of single vibrations is referred to as the spectrum determination or the spectrum (frequency) analysis of noise.

A transmission of the disturbing energy is connected with an acoustic wave propagation in a medium. The acoustic wave energy is characterized by the following notions and values:

- Sound power of a source, N_a, in W is the measure of the energy radiated by the source in the unit of time:

$$N_a = \frac{E_a}{t} \tag{6}$$

where E_a is acoustic energy of a source (W), and t is time (s).

- Sound intensity, I_a in W/m^2, the value of an sound power transmitted by a unit area perpendicular to the direction of an acoustic wave propagation:

$$I_a = \frac{N_a}{S} \tag{7}$$

where N_a is sound power (W), and S is surface area (m^2). The following relationship exists between the sound intensity, $I_a\left(W/m^2\right)$, and the acoustic pressure (p):

$$I_a = \frac{p^2}{p \cdot c} \tag{8}$$

where p is medium density $\left(kg/m^3\right)$, c is sound velocity (m/s).

Similarly to the acoustic pressure, because of the wide range of an acoustic power and sound intensity variations, the logarithmic scale is in use and notions of:

- Sound power level, L_N (dB):

$$L_X = 10\lg\frac{N_a}{10^{-12}} \tag{9}$$

- Sound intensity level, L_L (dB):

$$L_i = 10lg\frac{I_a}{10^{-12}} \tag{10}$$

The sound power level is the basic value characterizing the emission of noise from its source, and therefore it is applied for the estimation of machinery noisiness. In practice, the sound power level corrected according to the frequency-weighting characteristics A, known as sound power level A, is most often used.

The type of noise influence on people can be either annoying noise, not causing any permanent effects in humans, or harmful noise, which either causes permanent effects or creates a certain risk of its occurrence.

In respect of its time dependency, the noise is either steady, when the sound A level in a certain point varies not more than by 5 dB, or transient (variable in time, interrupted), when the sound A level in a certain point varies by more than 5 dB. One kind of transient noise is impulse noise, which consists of one or more sounds, each of a duration $< 1s\left(PN-N-01307:1994\right)$.

Taking into account its frequency range, noise can be divided into three groups:

- Infrasound noise, the spectrum of which contains components of infrasound frequencies from 2 to 16 Hz and of audible ones up to 50 Hz(PN – N – 01338:1986) [26].
- Audible noise, the spectrum of which contains components of audible frequencies from 16 to 16000 Hz.
- Ultrasound noise, the spectrum of which contains components of audible and ultrasound frequencies from 10 to 100 kHz.

There are also some other methods of differentiations of noise, e.g. taking into consideration the cause of its formation and the classification of its sources. There is, for example, aerodynamic noise – formed as a result of the movement of air or other gas, and mechanical noise, caused by friction and collision of solids, mainly elements of machines. There is also a way of dividing noise by considering the environment in which it occurs. A noise in industry is called an industrial noise, a noise in housing areas, public utilities, and rest areas is called a communal noise, while the noise in means of transportation a traffic noise.

The schematic presentation of the sound division when taking into account their frequencies is given in Figure 8.2.

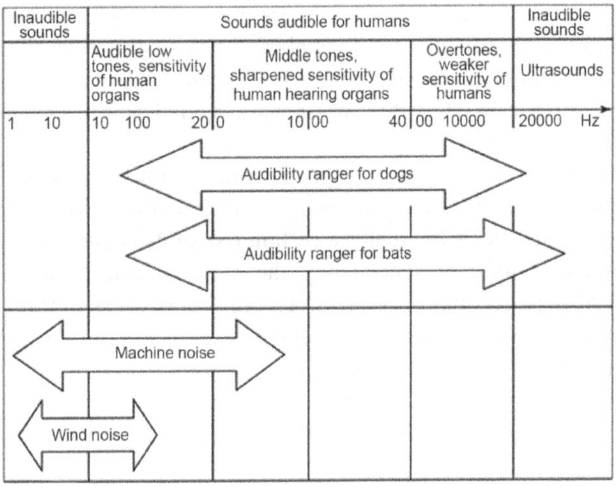

FIGURE 8.2 Division of sounds considering their frequency.

8.1.1 Loudness and Noisiness

The relation between physical values characterizing sound and a subjective feeling is not simple, however strictly defined. Subjective characteristic values of a sound are loudness, regarding its intensity, and timbre, regarding the spectrum frequency and timing (duration, time variation of the intensity – sound building up and decay).

Interrelations of sound loudness and intensity in the entire audibility range are presented in Figure 8.3. Narrow lines represent curves of the equal loudness level, giving the isophone contours. The intensity range is very broad, for example for the tone of 1000 Hz (standard value in acoustics), it spreads out from the threshold value of $2\times10^{-12}\,W/m^2$, hardly noticeable by an ear (equivalent to the threshold of audibility of an acoustic pressure of $2\times10^{-5}\,Pa$), to $10W/m^2$, being the threshold of pain.

Taking into account the wide span of the intensity of an auditory sensation area as well as the fact that the impression of loudness is (in agreement with Weber-Fechner's law – valid for all senses) proportional to the logarithm of the sound intensity, the level of the sound intensity (equation 40) related to the threshold of audibility of the 1000 Hz frequency tone is introduced: $I_0 = 10^{-12}\,W/m^2$.

In Figure 8.3, the straight horizontal lines represent equal sound levels from 0 to 130 dB in 10 – dB intervals (marked on the axis of ordinates), and

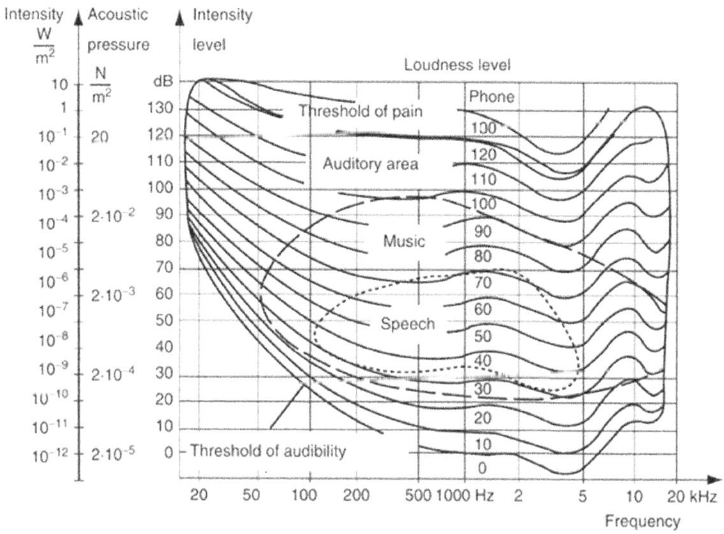

FIGURE 8.3 Interrelations of the sound loudness and intensity.

acoustic pressures are marked on the left-hand side of the same axis. The frequency given on the axis of abscissae is also in the logarithmic scale.

8.2 INFLUENCE OF NOISE AND ITS EFFECTS ON A HUMAN BODY

The negative influence of noise on a human body under the condition of occupational risk can be divided into two kinds (Figure 8.4):

- Health effects (hearing organ, general health condition, etc.)
- Functional effects (non-hearing-related influence on a human body)

8.2.1 Influence of Noise on a Hearing Organ

The harmful influence of noise on a hearing organ is connected with the following features and risk occurrences:

- A-weighted sound level or sound A level (for a steady noise) ⩾80 dB. Weaker stimuli are not harmful for the hearing organ even when acting for a long time without any interruption.

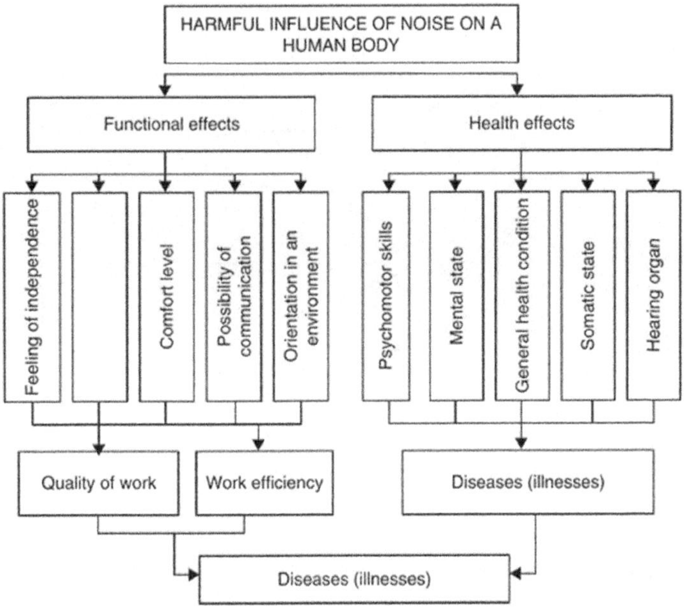

FIGURE 8.4 Negative influence of noise on a human body under the condition of occupational risk.

- Long time of the noise exposure. The effects accumulate in time, they depend on the acoustic energy dose transferred into a body in a certain period.
- Continuous exposure to noise is more harmful than interrupted. Even short breaks allow for the regeneration process of the hearing organ. The impulse noise is especially harmful. It is characterized by such rapid increase of acoustic pressure that the defending mechanism of the hearing organ preventing energy penetration into the ear has no chance to react.
- Noise spectrum with the majority of components being medium and high frequencies. The noise of such a spectrum is more dangerous for hearing than one with maximal energy in the low-frequency range. This is connected with the fact that the sensitivity of a human ear is highest in the frequency range 3 – 5kHz.
- Special, individual susceptibility to impairment of hearing. This depends on inherited features as well as acquired ones (e.g. as a result of past diseases).

The effects of noise on the hearing organ are as follows:

- Impairment of anatomic structures of the hearing organ (perforations, diminution of eardrum membrane) being mainly the result of a single and short exposure for noise at peak acoustic pressure >130 – 140 dB.
- Impairment of a hearing efficiency manifesting itself as the increase of the threshold of audibility. This is normally the result of a prolonged exposure to noise in the A-weighted sound level > 80 dB. This increase of the threshold can be reversible (temporary threshold shift) or permanent (permanent hearing loss).
- Audiometric examinations have revealed permanent hearing loss. An average permanent hearing loss of 30 dB at frequencies of 1000, 2000, and 4000 Hz on the side of the better ear, after making allowances for age, is considered the critical loss. This is the basis for diagnosis and medical certification of occupational deafness as the occupational disease.
- Occupational hearing loss (the occupational deafness), a permanent disability without the possibility for any improvement, has been for years ranked high on the list of occupational diseases. There are ~ 3000 new cases yearly, which constitute one-third of all cases registered in Poland.

8.2.2 Non-Hearing-Related Influence of Noise

The non-hearing-related influence of noise on a human body is not fully recognizable. The anatomical connection of a nervous hearing system with a cortex enables hearing stimuli to influence other brain centers (especially the central nervous system and the endocrine glands system), and as a consequence the state and functioning of many internal organs.

It has been experimentally proven that significant disorders of physiological bodily functions can occur when the acoustic pressure level > 75 dB. Weaker acoustic stimuli (55 – 75 dB) can cause distraction, hinder work, and reduce its efficiency. Non-hearing-related effects are generalized responses of the human body to sound exposure, and contribute to the progress of several illnesses (high blood pressure, gastric ulcers, neurosis, etc.).

The non-hearing-related effect of not being able to perceive and understand speech and sound security signals should be mentioned. Difficulties in verbal communication in a noisy environment (of 80 – 90 dB level) and the inability to perceive warning signals not only increase the difficulty of working conditions and reduce work efficiency but can also be reasons for injury at work. Being able to understand speech is one of the main criteria when evaluating noise in an environment.

8.3 MEASUREMENTS AND THE ESTIMATION OF NOISE IN AN ENVIRONMENT

Taking into consideration the purpose of measurements, there are two main methods to evaluate noise:

- Methods that measure machine noise
- Methods that measure noise in places occupied by people (in work environments)

8.3.1 Methods of Measuring Machine Noise

These methods are applied for the determination of values characterizing the noise of a machine considered as a separate noise source at certain experimental and operational conditions. Such values are: an acoustic power level A or an acoustic power level in frequency bands (calculated either on the basis of the acoustic pressure level or the sound intensity); the A-weighted sound level and the sound C peak level at workplaces (for non-portable machines in the workplace); a noise source direction indicator (if necessary); an acoustic pressure level; and a sound A level – measured at specified measuring points (e.g. where the acoustic power level is impossible to be measured). (According to standard EN ISO series 11200, this is known as the time-averaged sound pressure level of an acoustic emission corrected by characteristics A, and the peak level of an acoustic emission corrected by characteristics C).

The measured values are applied to:

- Compare measured levels with the established boundary values.
- Design, implement, and estimate noise control and reduction means.
- Compare machine noises either of the same kind or of different kinds.
- Establish the sound A level or the acoustic pressure level at a specified distance from the source.

All these values should be determined according to the European and international standards EN ISO 3740 and EN ISO 11200.

8.3.2 Methods of Measuring and Estimating Noise in Places Occupied by People

These methods are used to establish the degree of risk of being noise exposed for people at their workplaces and in specified distances from the noise source or sources. The results are mainly used for the comparison of existing acoustic conditions with the conditions required by standards and health-related regulations. They are also helpful in the estimation and choice of proper methods of noise control.

Methods of measuring noise in the work environment are given in standards. The instrumentation system, noise dosimeters, or integrating sound level meters of type 2, or better accuracy, and at least 53 dB impulse range should be applied (they should meet the requirements of IEC 804 and IEC 1252).

There are two basic methods of measurements: direct and indirect.

The direct method consists of continuous measurement during the whole time of worker exposure and of reading the meters, e.g. the noise dosimeter or the integrating sound level meter. The results will precisely describe the workers' risk.

The indirect method is based on noise measurement done in a shorter time than the one being assessed and the use of mathematical formulae for the estimation of values needed.

The mode and frequency of taking measurements, and the way they are registered, stored, and made available for workers are given in the regulations of the Ministry of Health and Social Security.

Estimation of occupational risk is based mainly on the comparison of measured or estimated noise values with permissible ones (the level of noise exposure, the maximal sound A level, and the peak sound C level) being in force simultaneously. Even if only one of the values is higher than allowed, the noise is regarded as exceeding the permissible value.

8.3.3 Permissible Noise Values in Work Environment

Noise values that are acceptable with regard to protecting hearing, according to the standards of many countries, are as follows:

- Noise exposure level related to an 8-h work day ($L_{EX,8h}$) should not exceed 85 dB and the equivalent daily exposure should not exceed 3.64×10^3 Pa^2 s or only when the noise is affecting a human body unevenly in separate days of a week.
- Noise exposure level related to the work week ($L_{EX,w}$) should be < 85 dB, and the equivalent week exposure should be 18.2×10^3 Pa^2 s.
- Maximum sound A level (L_{Amax}) should not exceed 115 dB.
- Peak sound C level $\left(L_{Cpeak}\right)$ should not exceed 135 dB.
- The last two values are important for the estimation of short and impulse noises.
- The standard regulations are compulsory unless there are other specific rules with yet lower values for certain demographics (e.g. a workplace with minors, $L_{EX,8h}$ 80 dB, or a workplace with pregnant women $-L_{EX,8h}$ 65 dB).
- Considering the requirement for workers to be able to fulfill basic tasks (this means considering non-hearing-related noise effects – the fatigue criterion), noise values should not be higher than those listed in Table 8.1.

TABLE 8.1 A-Weighted Sound Level $L_{Aeq,Tc}$ for Employees Present at Their Workplaces (in Accordance with PN-N-01307:1994)

NO.	*WORKPLACE*	*A-WEIGHTED SOUND LEVEL, $L_{Aeq,Tc}$, dB*
1	Direct control cabins without a phone connection, laboratories with noise sources, rooms with computers, typing machines, or fax machines, or other rooms of similar purpose	75
2	Dispatcher's cabins, observation and remote control cabins with a phone connection used in a process control, rooms for precision works, and other rooms of similar purpose	65
3	Administration rooms, design offices, rooms for theoretical works, data processing, and other rooms of similar purpose	55

Notice: Values in Table 8.2 do not relate to a 8-h work day. They concern nominal time, T_e, during which an employee is present at their specific workplace.

8.3.4 Noise Risk and Sound Sources in Work Environment

In many countries, large numbers of employees in various fields within industry are working under risk conditions because of annoying and stressful aspects of their work environments, including noise exposure levels of > 85 dB.

Those most at risk are workers of industry, building, forestry and transport, metallurgy, machine industry, chemical industry, mining, and electric power plants.

The main noise sources in machinery and equipment are shown in Figure 8.5.

Assuming that the main noise sources at workplaces are machines, equipment, and technological processes, one can specify the basic groups of noise sources [5]:

- Machines being the energy sources, e.g. engines (maximum sound A level up to 125 dB), compressors (up to 113 dB).
- Tools and pneumatic engines, e.g. hand operated pneumatic tools: hammers, chisels, grinders (up to 134 dB).
- Machines for comminution, crushing, sieving, cutting, cleaning, e.g. ball mills (up to 120 dB), vibration sieves (up to 119 dB), crushers

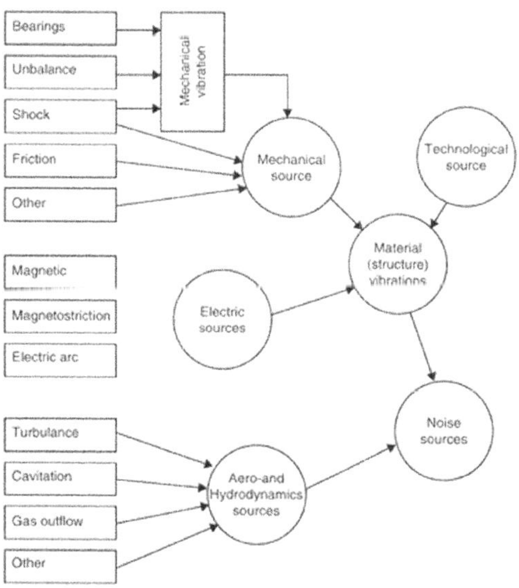

FIGURE 8.5 Noise sources in machines and equipment.

(up to 119 dB), shaking grids (up to 115 dB), circular saws for metals (up to 115 dB).

- Machines for plastic forming, e.g. pneumatic hammers (up to 122 dB), presses (up to 115 dB).
- Metal working machines, e.g. grinders, automatic lathes, boring machines (up to 104 dB).
- Woodworking machines, e.g. chisels (up to 108 dB), planers (up to 101 dB), molders (up to 101 dB), circular saws (up to 99 dB).
- Textile machines, e.g. winders (up to 114 dB), looms (up to 112 dB), spinning frames (up to 110 dB), drawing frames (up to 104 dB), twisters (up to 104 dB), carding machines (up to 102 dB), flow devices, e.g. valves (up to 120 dB), vents (up to 114 dB), works transport devices, e.g. overhead cranes, conveyers, feeders (up to 112 dB).

8.4 METHODS AND MEANS OF NOISE PROTECTION

According to European and national regulations, an employer is obliged to provide protection against hazards connected with noise exposure and to ensure the application of:

- Technological processes not generating any excessive noise.
- Machines and other technical equipment causing as quiet a noise as possible, not in excess of the quantity allowed.
- Solutions lowering the noise level at workplaces (with the priority for means of the noise reduction at its source).

At workplaces where, regardless of applying all possible means of noise reduction, the noise level still exceeds the limit, an employer should:

- Find out the reason of such noise and design a program of technological and organizational processes aimed at lowering that noise.
- Provide employees with hearing protection, matching the noise characteristics as well as individual features of workers, and enforce the use of such protection.
- Limit noise exposure time and introduce work breaks.
- Mark the noise hazard zones and, when reasonable and feasible, limit access to such zones by fencing them off. Persons employed at these workplaces must be provided with information concerning:

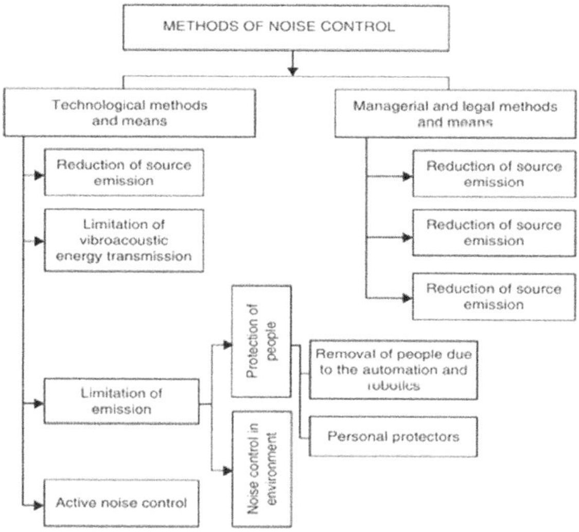

FIGURE 8.6 Methods of noise control.

- Noise measurements results and the health hazard they are facing.
- Processes being undertaken in connection with lower the noise that exceeds the limit.
- Proper selection of individual hearing protection devices and the way of their use.

When noise at workplaces exceeds the limit, employees undergo periodic medical examinations. In the case of significant and dynamic progress of hearing defects, the frequency of audiometric examinations should be increased and the time between successive tests shortened to 12 or even 6 months. In the case of an impulse noise hazard or noise of A-weighted sound level >110 dB, audiometric examination must be done at least once a year.

Methods of noise control are schematically presented in Figure 8.6.

8.4.1 Technical Means of Noise Reduction

8.4.1.1 Change from a Noisy Technological Process to a Less Noisy One

The loudest technological processes can be substituted by less loud ones, e.g. forging by hammer can be substituted by rolling, and forming while machining with hand-operated tools can be substituted by electric and chemical treatment and mechanized tools.

8.4.1.2 Mechanization and Automation of Technological Processes

Mechanization and automation of technological processes together with sound-insulated control cabins for personnel is one of the most modern, futuristic, and, at the same time, efficient way to eliminate noise and other hazards (e.g. dustiness, high temperature, injuries).

The majority of cabins used nowadays in industry ensures a noise reduction of 20 – 50 dB at frequencies > 500 Hz.

8.4.1.3 Constructing and Implementing Silent-Running Machines, Equipment, and Tools

Changes to technological processes and the introduction of mechanization and automation require longer periods of realization and cannot be implemented for small-series or atypical production.

Very efficient silencing of sound sources in machines (the reduction of a sound emission) can be done by:

- Reduction of forces causing vibrations (and restricting their spectra), e.g. by careful balancing of machine elements, changing stiffness and system structures, changing frictional resistance.
- Changing aerodynamic and hydrodynamic conditions (e.g. changing geometry of intake and exhaust of energetic media and a velocity of their flow).
- Reduction of a radiation efficiency coefficient (e.g. changing dimensions of vibro-acoustic energy radiating elements, changing materials, and installing insulation of boards in a system).

8.4.1.4 Acoustically Correct Layout of a Plant and Utilization of Rooms

The following guidelines should be observed at the design stage of industrial buildings:

- Buildings and rooms requiring silence (e.g. laboratories, design offices, places for the scientific research) should be isolated from buildings where a noisy technological process is being performed.

- Machines and equipment should be grouped, if feasible, in separate rooms, according to their loudness.

Noise can significantly increase by improper utilization of a room, especially when machines are too densely located. The shortest recommended distance between machines is 2 – 3 m.

8.4.1.5 Acoustic Dampers

Reduction of noisiness in the air and other gas ducts (ventilation systems, intake and exhaust systems of fluid-flow machines, e.g. compressors, blast machines, turbines, engines) can be achieved by the application of acoustic dampers. Modern construction of acoustic dampers do not cause any losses of a machine power. They create a high resistance against noisy transient flows while not choking steady flows that transport the air or gases. A well-known damper of this kind is the reflexive damper, which contains acoustic wave filters with absorptive dampers that contain the sound-absorbing material.

Reflexive dampers operate on the basis of the reflection and interference of acoustic waves and have very good damping qualities in the range of low and medium frequencies. They are applied where the flow velocities are significant and the temperatures high, mainly in engines, blowers, compressors, and sometimes also in ventilators.

The absorption dampers counteract the acoustic energy transmission along ducts by absorbing the majority of that energy by the sound-absorbing material. They damp mainly medium and high frequencies and therefore are used in ventilation ducts. In practice, both kinds of dampers are often used because industrial noise sources emit the energy in a wide range of frequencies.

A different kind of damper is the active damper. In Figure 8.7, the classification of dampers together with their schematic presentation is shown.

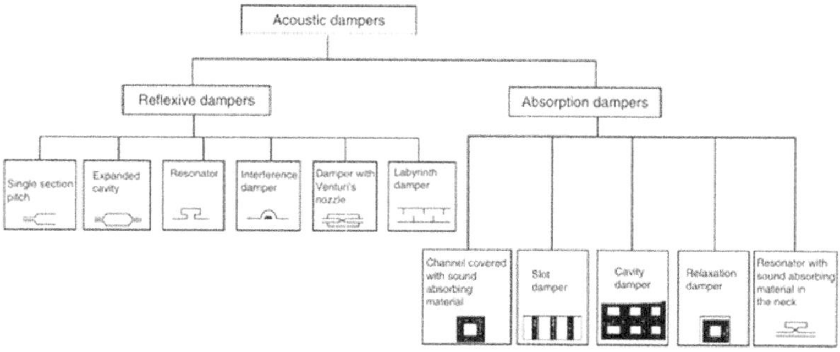

FIGURE 8.7 Classification of dampers.

8.4.1.6 Sound-Insulated Enclosures

An attenuation of noise can be achieved by enveloping the noise source, as a whole or a part of it, in an enclosure. Sound-insulated enclosures should efficiently damp sound waves emitted by the sound source while not disturbing the normal work and maintenance of enveloped machines.

Typical, the most often used enclosures have sound-insulated walls made out of steel sheets covered from the inside by a damping mass or sound-insulating materials. There are also enclosures with multilayered walls.

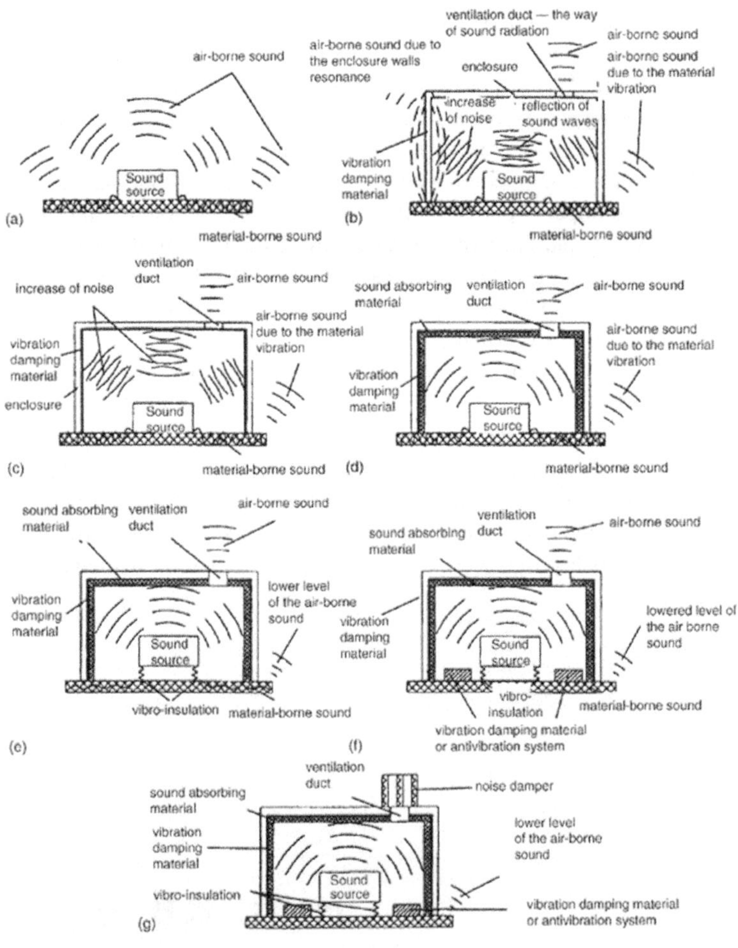

FIGURE 8.8 Principle of operation of the enclosure.

Correctly made enclosures can attenuate the sound A level by 15 – 20 dB. Partial enclosures are much less efficient.

Ventilation ducts and other openings, necessary for the technological process, lower the performance of the enclosure. Application of acoustic silencers, e.g. in the form of channels covered with the sound-absorbing material, is needed for openings.

Examples illustrating the principle of operation of the enclosure are given in Figure 8.8.

8.4.1.7 Sound-Absorbing Screens

Sound-absorbing screens are used as shields in workplaces. Their aim is to damp the noise emitted by different machines in the direction of a workplace as well as to damp the emission from that place into an outside space. To achieve the maximum efficiency, the screen should be placed as close as possible to the sound source or the workplace.

Basic elements of a screen are: an insulation layer inside (usually a steel plate of a proper thickness) and sound-absorbing layers outside (plates of mineral or glass wool shielded by perforated sheets).

When the screen is being used in the interior, it should be a part of the whole acoustic system to cooperate with other elements damping the reflected waves energy (sound-absorbing materials). Effectiveness of properly installed sound-absorbing screens is estimated to be 5 – 15 dB in the distance of 1.5 m behind the screen on the axis perpendicular to its surface.

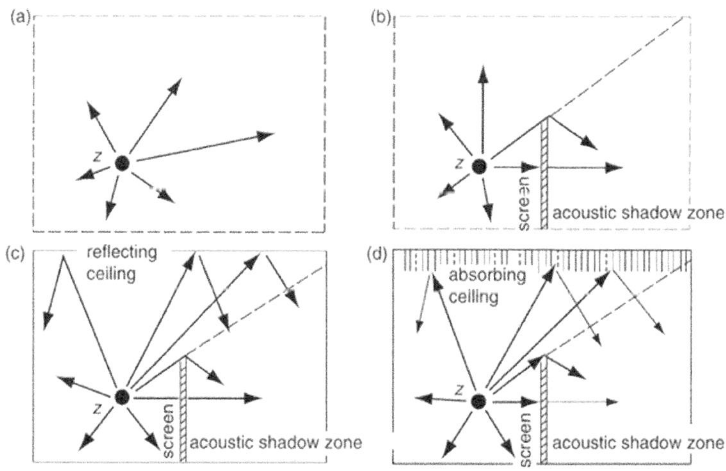

FIGURE 8.9 The principle of operation of the acoustic screen.

The principle of operation of the acoustic screen is shown in Figure 8.9. The acoustic energy source in an open space is shown in Figure 8.9 (a). The same source with the screen is seen in Figure 8.9 (b) (also in an open space). The acoustic energy source with the screen located inside the room with the ceiling reflecting acoustic waves is seen in Figure 8.9 (c), and absorbing them in Figure 8.9 (d).

8.4.1.8 Sound-Absorbing Materials

Sound-absorbing materials applied on walls and a ceiling of a room increase its absorbing capacity. Done in such a way, the lowering of the reflected waves sound level leads to the diminishing of the total noise level in the room.

Porous materials, such as textiles, mats of mineral and glass wool, plates and porous wall plasters, plates and porous plastic mats, plastics sprayed under pressure, are the most often applied.

The maximum of the adsorption coefficient of the properly chosen sound-absorbing material should occur in the same frequency range where there is the maximum component of the noise spectrum and the allowed noise values are exceeded.

As experience shows, satisfactory damping results (lowering the noise level by 3 – 7 dB) could be achieved in such rooms where the original noise absorption was small.

Sound-absorbing systems such as ceilings, partitions, and protective walls, produced in Poland or imported from abroad, are available nowadays off-the-shelf.

8.4.1.9 Hearing Protectors

In all cases when the sound level near the workplace exceeds the allowed value and no means of noise reduction is available, hearing protectors should immediately be used.

Such protectors should also be used when excessive noise does not occur often, or when a worker maintaining a noisy machine only occasionally has to enter the place where this machine is located.

The basic value describing the acoustic performance of hearing protectors is their damping of the sound (referred to previously as acoustic effectiveness) [8]. Hearing protectors are considered satisfactory when the A-weighted sound level under the protector is lower than the admissible value (85 dB).

Considering their construction, hearing protectors can be divided into: antinoise pads (expendable or reusable), antinoise ear-protectors (either with an on-head hold down spring or an on-helmet one), and anti-noise helmets.

Selection of protectors for each real acoustic condition should be done with the consideration if they are going to give the satisfactory protection to the hearing organ. Choosing protectors for each workplace is done on the basis of acoustic pressure levels in octave frequency bands or sound A and C levels measured at those places and parameters of hearing protectors granted the certificate of the safety mark B.

8.4.1.10 Active Methods of Noise Reduction

Low-frequency noise is especially difficult to limit. Known and used for years, traditional (passive) noise reduction methods for frequencies below 500 Hz are not effective and very costly. In recent years, more and more often active methods are being used. Their characteristic feature is the compensation of noise by sounds from the additional external energy sources. These sources are controlled by signals of the specially shaped amplitudes and phases. Matching the control factors, it is possible either to provide or to absorb the vibro-acoustic energy from specified places in the system.

The general principle of the active compensation of the acoustic field parameters is as follows:

- Primary sound source, called the compensated source, generates the acoustic wave, called the primary wave or the compensated wave.
- Secondary sound source, called the compensating source, generates the secondary acoustic wave, called the compensating wave.
- Destructive interference of both waves takes place at the specified observation point.

To get a satisfactory noise reduction, strictly described relations between amplitudes and phase shifts of the compensated and the compensating signals must occur at the observation point. In an ideal case, the full reduction of the compensated signal will be reached. It will happen only when the compensating signal is the ideal inversion of the compensated one.

Active noise reduction systems, most often applied in practice, are active noise dampers of fluid-flow machines and internal combustion engines (the damping efficiency is 15 – 30 dB for the frequency up to 600 Hz). Active systems are also used often in hearing protectors; they allow damping effectiveness to increase by 10 – 15 dB in the frequency range 50 – 300 Hz.

8.5 INFRASOUND NOISE

8.5.1 Factor Characteristics

When the spectrum of noise contains components of infrasound frequencies (2–16 Hz) and audible frequencies (up to 50 Hz), this is known as infrasound noise. At present, another expression – low-frequency noise (comprising the range from 10 to 250 Hz) – is more and more often used.

Infrasounds, contrary to popular belief of their inaudibility, are received by the body by the manner (mainly by the hearing organ). Their audibility depends on the acoustic pressure level. The individual hearing perception especially of low-frequency sounds is very different. The lower the frequency, the higher their audibility threshold, e.g. for frequencies 6 – 8 Hz, the threshold is at ~100 dB, for frequencies 12 – 16 Hz at ~90 dB.

Apart from the manner, infrasounds are also received by receptors of vibration. Thresholds of this perception are ~20 – 30 dB higher than audibility ones.

When the acoustic pressure level >140 dB, infrasounds can cause permanent harmful changes in the body. The resonance of internal organs of the human body can occur, which is subjectively felt as an unpleasant internal vibration (from 100 dB already). Apart from heaviness in ears, this is one of the most typical symptoms identified by persons under the influence of infrasounds. However, the dominant effect of infrasounds, occurring even at small excesses of the audibility threshold, is fatigue. Over-tiredness, discomfort, sleepiness, losing balance and psychomotor skills, and disorders of physiological functions are symptoms sensed to different degrees by individuals. Changes in the central nervous system characteristic of a decrease in vigilance (especially dangerous for drivers and machine operators) are the objective proof of these subjective feelings.

The main infrasound sources in a work environment are fluid-flow low-rotation machines (compressors, ventilators, engines), electrical equipment (mills, boilers, chimneys), metallurgical furnaces (especially electric-arch furnaces), and casting devices (molding machines, shakeout grits).

8.5.2 Permissible Values and Measuring Methods

Infrasound noise at workplaces is characterized by the acoustic pressure level in octave bands with mid-band frequencies of 8, 16, and 31.5 Hz. The

TABLE 8.2 Permissible Values of the Infrasound Noise Established for the Health Protection of Employees

MID-BAND FREQUENCY OF OCTAVE BANDS (Hz)	*ACOUSTIC PRESSURE LEVEL FOR 8 H NOISE EXPOSURE (dB)*	*MAXIMAL ACOUSTIC PRESSURE LEVEL (dB)*
8; 16	110	137
31.5	105	132

TABLE 8.3 Permissible Values of Infrasound Noise Workplaces of Minors and Pregnant Women

MID-BAND FREQUENCY OF OCTAVE BANDS (Hz)	*ACOUSTIC PRESSURE LEVEL FOR 8H NOISE EXPOSURE (dB)*
8;16	85
31.5	80

TABLE 8.4 Permissible Values of the Infrasound Noise – with the Ability to Perform Basic Tasks by Employees Taken into Account

NO.	*WORKPLACE*	*MID-BAND FREQUENCY OF OCTAVE BANDS (Hz)*	*ACOUSTIC PRESSURE LEVEL FOR 8 H NOISE EXPOSURE (Hz)*
1	Dispatcher's cabins, remote control cabins, rooms for precision works and others of similar purpose	4;8;16 31.5	90 85
2	Administration rooms, design offices, rooms for theoretical work, data processing and others of similar purpose	4;8;16 31.5	85 80

permissible values established for the health protection of employees are listed in Table 8.2.

In protected workplaces of minors and pregnant women, the permissible values are lower (Table 8.3) [18,19].

Considering the employees' ability to perform basic tasks (the fatigue criterion of infrasound), acoustic pressure levels should not exceed values given in Table 8.4.

Methods of infrasound noise measuring are described in normatives. Measurements can be done in a direct way (utilizing a mobile frequency analyzer) or in an indirect way (when noise is measured at industrial conditions and the frequency analyses done in a laboratory).

Presently, in accordance with the international standard ISO 7196:1995 [34], measurements of a sound G level, which refers to when the acoustic pressure level is corrected to the frequency of G characteristics, are also recommended. G characteristics correspond roughly to the threshold of infrasound hearing perception and – as the experimental results show – correlate well with the subjective estimation of the sound intensity.

According to recommendations included in the newly established standards ISO 9612:2009 and ISO 7196:1995, and proposals by the American Conference of Government Industrial Hygienists (ACGIH), as well as with regard to an increase in the knowledge of infrasound influence on the human body and the development of measuring techniques, discussions on the verification of national NDN infrasound noise standards are planned.

8.5.3 Methods of Hazard Limitations

In prevention of the harmful influence of infrasound noise, the same rules and regulations as with normal noise apply. However, the protection against infrasounds is more complicated because of the significant length of the infrasound waves (20 – 170 m) for which traditional walls, partitions, screens, and sound absorbers are not very effective. In some cases, infrasound waves are increased because of the resonance of rooms, structural elements, and whole buildings.

The best protection against harmful infrasound noise is its suppression at the generating source, referring directly to machines and devices.

Other solutions are:

- Use of sound dampers at inlets and outlets of the air (or gas) in fluid-flow machines.
- Proper foundation engineering (with vibroinsulation) of machines and devices.
- Stiffening of walls and building structures to prevent resonance.
- Applying sound-insulated cabins of a heavy structure (made of bricks) for machine and installation operators.
- Applying active methods of noise reduction (connected with the active absorption and sound compensation).

8.6 ULTRASOUND NOISE

8.6.1 Factor Characteristics

An ultrasound noise is a noise where the spectrum contains components of audible and ultrasound frequencies from 10 to 100 kHz.

Air ultrasounds, being a part of the ultrasound noise, can penetrate the human body through the hearing organ as well as the whole body surface. Investigations of the influence of ultrasound noise on the hearing organ are rendered difficult because in an industrial environment, ultrasounds are usually accompanied by audible noise. Then it is difficult to assess if the deterioration of hearing is the result of either the audible or the ultrasound components or simultaneous action of both types.

However, the prevalent opinion nowadays is that as a result of non-linear phenomena occurring inside the ear, the subharmonic components are formed under the influence of the ultrasounds. Their acoustic pressure level is of the same order of magnitude as the basic ultrasound partial. In consequence of that phenomenon, hearing impairment concerns the subharmonic frequencies of the ultrasounds.

Apart from harming hearing, a negative influence on the vestibular organ in the internal ear, manifested by headaches and dizziness, balance disorders, nausea, drowsiness, and over-tiredness, has been found.

Examinations of non-hearing-related influences have shown that occupational exposure of ultrasound noise of levels > 80 dB in the range of high audible frequencies, and > 100 dB in the range of low ultrasound frequencies, causes changes of a vasovegal nature.

The main sources of ultrasound noise in a work environment are technical ultrasound devices, which generate ultrasounds indispensable for performing technological processes. Ultrasound noise can be also generated by high-speed engines: compressors, pneumatic devices, metal working machines, and some textile machines.

8.6.2 Permissible Values and Measurements Methods

Ultrasound noise at workplaces is described by acoustic pressure levels in one-third-octave bands of mid-frequencies from 10 to 100 kHz.

TABLE 8.5 Values of the Acoustic Pressure Level of Ultrasound Noise Allowed Considering Employees' Health

MID-BAND FREQUENCIES OF ONE-THIRD-OCTAVE BANDS KHz	*ACOUSTIC PRESSURE LEVEL FOR 8 HR NOISE EXPOSURE dB*	*MAXIMAL ACOUSTIC PRESSURE LEVEL dB*
10;12.5;16	80	100
20	90	110
25	105	125
31.5;40;50;63;80;100	110	130

TABLE 8.6 Permissible Values of the Acoustic Level at Workplaces of Minors and Pregnant Women

MID-BAND FREQUENCIES OF ONE-THIRD-OCTAVE BANDS KHz	*ACOUSTIC PRESSURE LEVEL FOR 8 HR NOISE EXPOSURE dB*	*MAXIMAL ACOUSTIC PRESSURE LEVEL dB*
10;12.5;16	75	77
20	85	87
25	100	102
31.5;40;50;63;80;100	105	107

Permissible values, established with a view to protect employees' health, cannot surpass the values given in Table 8.5.

At protected workplaces (those with minors or pregnant women), lower values are required. These are presented in Table 8.6.

8.6.3 Methods of Risk Limitation

The same rules and requirements apply for the prevention of the harmful influence of ultrasound noise as for normal noise. Medical examination of workers should be performed every two years.

Taking into consideration the propagation of short-waves (of low-frequency ultrasounds) in the air (wavelengths from 3 mm to 2 cm), it is relatively easy to reduce its harmful influence on humans, e.g. by airtight sealing and enclosing of sources, remote control of technological processes in which ultrasounds are used, avoiding direct contact with an ultrasound transducer, using individual protection means, etc.

BIBLIOGRAPHY

American National Standard ANSI S 12.12.1002 (1992).

Beranek, L.L. and Ver, I.L. (1991). *Noise and Vibration Control Engineering: Principles and Applications*. New York: Wiley.

Commission Proposal for a Council Directive on the Minimum Health and Safety Requirements Regarding the Exposure of Workers to the Risk arising from Physical Agents, 93/C77/01, OJ No. C77, 1993, p. 11, 94/C130/ 03, OJ No. C130/3, 1994.

Crocker, M.J. (ed.). (1997). *Encyclopaedia of Acoustics*. New York: Wiley.

EN ISO 11200. Acoustics - Noise Emitted by Machinery and Equipment.

Engel, Z. (1993). *Noise and Vibration Control in Environment*. Warsaw: PWN. [in Polish].

Fasold, W., Kraak, W., & Schirmer, W. (1984). *Taschenbuch Akustik*. Bd. I, II. Berlin: VEB. [in German].

Harris, C.M. (ed.) (1979). *Handbook of Noise Control*. New York: McGraw-Hill.

IEC 804: 1998 (1998). Integrating - Averaging Sound Level Meters.

IEC 1252 Electroacoustics - Specifications for Personal Sound Exposure Meters.

FURTHER READING

Mir, M., Nasirzadeh, F., Bereznicki, H., Enticott, P., Lee, S., & Mills, A. (2023). Construction noise effects on human health: Evidence from physiological measures. *Sustainable Cities and Society*, 91, 104470.

Mehrotra, A., Shukla, S. P., Shukla, A. K., Manar, M. K., Singh, S. K., & Mehrotra, M. (2024). A comprehensive review of auditory and non-auditory effects of noise on human health. *Noise and Health*, 26(121), 59–69.

Patel, D. B., & Solanki, H. K. A. (2021). Effects of noise pollution on human health. *Research and Reviews: Journal of Environmental Sciences*, 3(1), 1–5.

Thompson, R., Smith, R. B., Karim, Y. B., Shen, C., Drummond, K., Teng, C., & Toledano, M. B. (2022). Noise pollution and human cognition: An updated systematic review and meta-analysis of recent evidence. *Environment International*, 158, 106905.

Yavuz, A., Hazar-Yavuz, A. N., & Hacibektaşoğlu, S. E. (2024). Vehicle noise pollution awareness for human health and environmental impacts: A comprehensive review. *ACTA Pharmaceutica Sciencia*, 62(2), 244–266.

9 Noise at Work

P. J. School

9.1 INTRODUCTION

Undesirable sound in the 20 Hz to 20000 Hz(20kHz) range is called noise. Noise may be constant or may vary in duration, frequency, or magnitude. Kryter (1970) and Price et al. (1989) provide convincing evidence of the debilitating effects of noise on humans. Noise impairs human performance, and high levels cause temporary hearing impairment or permanent hearing loss. Therefore, noise is a human factors engineering consideration.

The following is a succinct description of essential concepts that may be used to quiet freestanding items such as pumps, motors, or engine-powered equipment. A very extensive coverage of noise and noise control is provided by Harris (1991). Lord et al. (1980) is a readable text that contains several examples of practical designs. Hirshorn (1989) is a concise reference.

DOI: 10.1201/9781003675761-9

9.2 NOISE REDUCTION

9.2.1 Feasibility

Feasibility determination involves methods that permit the prediction of the magnitude of noise reduction and the cost of noise reduction effort. Basically, data are gathered on a predecessor or mock-up item. Those data are used to generate a notional design and a cost estimate to implement that design. That estimate is compared with medical, worker compensation, or other costs that would accrue if the item was not quieted.

If noise reduction is infeasible, measures such as posting warning notices requiring use of hearing protectors or prohibiting access to a noisy area may be used. Because most hearing protection devices impair speech intelligibility and some devices cause discomfort, personnel tend to avoid their use. Also, people occasionally forget to wear hearing protectors. Consequently, the effectiveness of such measures relates to training, monitoring, and management control.

9.2.2 Reductions Achieved

Over the past 15 years, quieting efforts on several freestanding, diesel engine-powered items produced 5–20 dB noise reductions. Those noise reductions led to attractive cost and human performance benefits.

This experience showed that 6 dB reductions are often readily achievable. Twelve-decibel reductions require more effort and insight. In a few cases, reductions of >18dB were achieved. These large noise reductions involved relocating noisy components, adding vibration isolation, adding parts such as intake and cooling air silencers, or balancing rotating parts.

9.2.3 Sound Level Measurement

Subjective judgement of sound levels is unreliable; therefore, noise assessment or abatement requires sound level measurement and first decimal place accuracy. Instruments that comply with the American National Standards Institute (ANSI) S1.4 Type 0 or Type I sound meter requirements provide sufficient accuracy (ANSI S1.4-1983, para. 3.2). A calibrated Type 0 meter is accurate to ±0.4 dB; the Type I is accurate to ±0.7 dB.

9.2.4 Fundamentals

9.2.4.1 Sound Levels

Sound levels in decibels may be obtained from sound power or sound pressure measurements. Sound power levels $\left(L_w\right)$ are expressed as acoustic power in watts (W). Sound pressure levels $\left(L_p\right)$ are expressed as the magnitude of air pressure changes in pascals. The term "sound level" indicates that the data are log base 10 transformed.

9.2.4.2 Sound Power Level

Sound power level in decibels (dBL_w) is defined as:

$$L_W = 10\log_{10}\left(W / W_{ref}\right),$$

where W_{ref} is 10^{-12} Watt or

$$L_W = 10\log_{10}W + 120.$$

Thus, sound power in decibels equals 10log based 10 of the ratio of a sound power level to the 10^{-12} Watt reference level.

9.2.4.3 Sound Pressure Level

Sound pressure level in decibels (dB L_p) is defined as:

$$L_p \text{ dB} = 10\log_{10}\left(P / P_{ref}\right)^2, \text{ where } P_{ref} \text{ is} 20^{-6} \text{ Pascals}$$

or

$$L_p \text{ dB} = 20\log_{10}\left(P / P_{ref}\right).$$

Thus, sound pressure level in decibels equals 20 log based 10 of the ratio of a sound pressure to the 20^{-6} pascals reference level.

9.2.4.4 Decibels Are Logarithmic

The sound level algorithms contain a $\log_{10}$ term; therefore, sound levels must be logarithmically added or subtracted. The following decibel addition and subtraction nomographs provide approximations. The algorithms that follow the nomographs are more accurate than the nomographs.

9.2.4.5 Decibel Addition

To add two sound decibel levels $(L_1 + L_2)$, subtract L_1 from L_2 to yield Xs; then arithmetically add the corresponding Y to the larger of the two levels.
Add

X =	0	1	2	3	4	5	6	7	8	9
Y =	3	2.5	2.1	1.8	1.5	1.2	1	0.8	0.6	0.5

Add

X =	10	11	12	13	14	15 to 20	20
Y =	0.4	0.3	0.3	0.2	0.2	1	0.04

Example: If L_1, L_2 are both 100 dB, the difference is 0; thus, $X = 0$ and $Y = 3$ dB; so, 100 dB + 100 dB = 103 dB.

The preceding nomograph is based on the following algorithms:

$$\log^{-1}(Lt/10) = \log^{-1}(L_1/10) + \log^{-1}(L_2/10)$$

or

$$L_t = 10\log_{10}\left[\log^{-1}L_1/10) + \log^{-1}(L_2/10)\right]$$

where

L_t = total sound level

L_1 = a sound level

L_2 = another sound level.

9.2.4.6 Decibel Subtraction

If one has a known overall sound level (L_t) which is the sum of two levels, a known level (L_1) and an unknown level (L_2), then L_2 can be determined by arithmetically subtracting L_1 from L_t to yield "X" for the following nomograph. Then, the "Y" that corresponds to the "X" may be subtracted arithmetically from L_t to yield the unknown level L_2.
Subtract

X = 0	1	2	3	4	5	6	7	8	9
Y = 16.4	6.9	4.3	3	2.2	1.7	1.3	1	0.8	0.6

Subtract

X =	10	11	12	13	14	15 to 19	20
Y =	0.5	0.4	0.3	0.2	0.2	0.1	0.04

Example: If the overall sound level (L_t) is 100 dB and one of the two emitters (L_1) is 95.0 dB, then $X = 5$ and $Y = 1.7$. Therefore, the L_2 level is 100 dB $- 1.7 = 98.3$ dB.

The decibel subtraction algorithms are

$$L_2 = 10 \quad \log_{10}\left[\log^{-1}(Lt/10) - \log^{-1}(L_1/10)\right.$$

$$\text{delta} = \log^{-1}(L_L/10) - \left\{10\log_{10}\left[\log^{-1}(L_t/10)\right.\right.$$

$$\left.\left.-\log^{-1}(L_1/10)\right]\right\},$$

where

L_t = total sound level

L_1 = measured sound level of one emitter

L_2 = an unknown sound level produced by another emitter

delta = the amount to be subtracted from the overall (L_t) level.

9.2.5 Inverse Distance Relationship

Sound waves from a point source that is surrounded by nothing but uniform density air (a free field) radiate in a spherical pattern. Under those circumstances, sound pressure levels are inversely proportional to the distances from the emitter.

A way to extrapolate sound pressure level over distance in a free field is called the "6 dB rule." According to the rule, sound pressure levels increase or decrease by 6 dB if the distance to the source is halved or doubled. Using the rule, if the sound pressure level emitted by a point source is 100 dB at 10 m, then the level at 5 m would be $100 + 6$ dB $= 106$ dB; and at 20 m, the level would be $100 - 6$ dB $= 94$ dB.

Sound waves radiating from a sizeable emitter do not become spherical until they travel some distance from the emitter. The area where sound waves have not become spherical is called "the near field." Extrapolations to or from the near field are likely to be inaccurate. Therefore, in the near field, sound level measurements are preferred to extrapolations. Textbooks suggest that the near field extends four to seven times the largest item dimension. Thus, if an item's largest dimension is 2 m, then the near field could extend $8-14$ m from the item.

9.2.6 Octave Data

An audio spectrum is needed to decide the feasibility of noise reduction and the design. Full octave or fractional octave measurements can provide a sound magnitude versus sound frequency spectrum.

Noise measurement octaves are approximate frequency multiples. Commonly used full octave band center frequencies are 31.5,62,128,250,500 Hz, etc. One-third octave noise measurements provide a more complete audio spectrum. The common one-third octave center frequencies are 31.5,40,50,63,80,100,125,160,200,250 Hz, etc.

9.2.7 Sound Level Weighting

A widely published family of curves (Fletcher and Munson 1933) shows that human hearing is non-linear with respect to both frequency and sound pressure. "A-Weighted Decibel Level" (dB(A)) measurements take the non-linearity of human hearing into account. Therefore, much noise reduction work is done with dB(A) data.

9.2.8 Sound Wave Fates

Though other phenomena may occasionally be of interest, three sound wave fates are almost always of interest. A sound pressure wave may

1. radiate in all directions from an emitter;
2. be reflected from a surface; or
3. be absorbed by a surface.

If a sound wave strikes a surface, some of it may be transmitted through the surface.

For simplicity, Figure 9.1 depicts a single set of wave paths as straight lines rather than a spherical pattern.

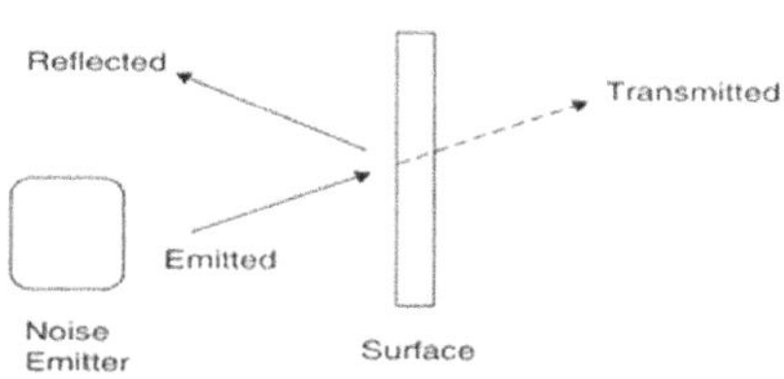

FIGURE 9.1 Single set of wave paths.

If a sound wave

1. does not strike a surface but instead radiates through uniform density air, the sound pressure level will decrease 6.02 dB with each doubling of the distance; or
2. if the wave strikes a mostly reflective surface, most of the sound will be reflected toward the emitter, some of the sound will be absorbed in the surface, and the rest will be transmitted through the reflective surface; or
3. if the wave strikes a mostly absorptive surface, most of the sound will be absorbed (converted to heat) in the surface, some of the sound will be reflected by the surface, and the rest will be transmitted through the absorptive surface.

Most surfaces reflect, absorb, and transmit sound. Designing a noise-reducing surface typically involves using those characteristics in ways that result in sufficient noise reduction while avoiding unacceptable cost, weight, or size gains.

9.2.8.1 Transmission Loss

Sound level reductions that occur from one side of a surface to the other are called "transmission loss." Reflecting sound or absorbing sound can produce transmission loss. The noise spectrum of interest and design limits (such as weight and size) determine whether one uses reflection, absorption, or the combination of both to achieve transmission loss.

9.3 NOISE-REDUCING AN ITEM

9.3.1 Noise Levels Add Logarithmically

The overall sound level produced by a group of emitters is the sum of the levels produced by each emitter in the group. For example, if an emitter had a sound level of 100 dB and a second emitter was added, then the sound level would increase to 103 dB. On the other hand, if two emitters are > 10 dB apart, the combined level will increase < 0.5 dB. Thus, noise-reduction effort usually focuses on emitters that are < 10 dB of the overall level.

9.3.2 General Noise-Reduction Strategies

Two common strategies for reducing noise are (1) surrounding the noisy item with a noise attenuating enclosure or (2) quieting emitters that comprise the item.

If one cannot change the design of the noisy item itself, noise reduction usually involves an enclosure. With engine-powered items, enclosure design usually requires the use of absorption and reflection. Many items require cooling airflow, and that may be provided by sound-attenuating ducts.

9.3.3 Noise Emitter Determination

Noise reduction efforts begin with overall noise and significant emitter noise measurements. To focus attention productively, one ranks significant emitters (emitters that are within 10 dB of the overall noise level) from highest level to lowest level. Significant emitters may be located by measuring levels at short distances, by making suspected components temporarily inoperative, or by touching suspected areas to detect vibrations or rattles. Suspected high frequency (> 1kHz) emitters may be confirmed by surrounding them with home insulation-type glass fiber. To confirm suspected low frequency (< 1kHz) emitters, surround them with lead-filled foam polymers or other dense material such as lead roof flashing.

9.3.4 The Ideal Mass Law

Thin, lightweight absorptive materials produce large transmission losses at frequencies > 1 or 2 kHz. At frequencies < 1kHz, absorptive materials are usually impractical because the thickness needed would be unacceptable. So, reflection is used to achieve low frequency transmission losses.

Reflection-caused transmission loss for a single wall surface can be estimated with an idealized approximation of "the mass law." Hirshorn (1989: D-2) provides the following algorithms for ideal transmission loss sound pressure calculations.

9.3.4.1 Ideal Mass Law for English Units

Transmission loss $\left(\mathrm{T_L}\right) = 20\log W + 20\log F - 33.5$,

where

W = mass density of the reflecting surface (lb per ft^2)

F = frequency (Hz).

9.3.4.2 *Ideal Mass Law for Metric Units*

Transmission loss $(T_L) = 20\log W + 20\log F - 47.5$,

where

W = mass density of the reflecting surface (kg per m^2), m^2), F = frequency (Hz).

Figure 9.2 shows ideal transmission loss for two surface mass densities. Please note that each doubling of either the mass density or the frequency of sound impinging on the surface increases the transmission loss 6 dB.

A comparison of ideal mass law calculations to data gathered with 10lb/ft^2 material shows that calculated values deviate from measured transmission losses by 5–8 dB at frequencies <1kHz (Hirshorn 1989: D-3). Low frequency deviations from ideal mass law numbers are caused by factors described in some detail by Lord et al. (1980; 242–247).

To adjust for ideal mass law deviations from empirical data, we estimate low frequency transmission losses by subtracting 6 dB from each of the calculated ideal transmission loss numbers <1kHz. Then, we use measurements made on prototyped enclosures to determine if the low frequency transmission losses lead to sufficient overall noise reduction. If not, we adjust the design by changing the surface density or by adding damping to the surface.

Damping increases low frequency transmission losses by converting sound to heat or by shifting the surface resonance to a lower frequency. Thus, damping can avoid some of the weight gain associated with increasing the surface mass density of an untreated surface. Damping may be achieved by adhesively attaching damping materials to the surface or by making the surface with composite materials that provide good damping.

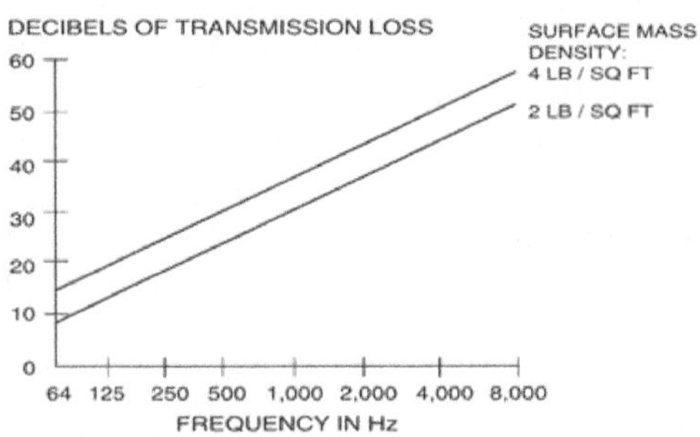

FIGURE 9.2 Ideal mass law transmission loss.

9.3.5 Calculating Absorption

Many acoustics texts list sound absorption coefficients for commonly used materials. Additionally, manufacturers of acoustic materials provide lists of absorption coefficients.

According to Hirshorn (1989; C-6), the sound absorption coefficient equals sound energy absorbed (converted to heat) divided by the incident sound energy. Therefore, absorption-caused transmission losses are calculated by multiplying the sound power level by the appropriate (frequency dependent) absorption coefficient of the absorptive material. To do that, sound pressure levels $\left(L_{\mathrm{p}}\right)$ are converted to sound power levels $\left(L_{w}\right)$ that are multiplied by the appropriate (frequency dependent) absorption coefficients.

According to Harris (1991; 1.12, 1.13), the metric units L_{w} to L_{p} to algorithm is:

$$L_{\mathrm{W}} = L_{\mathrm{p}} + 20\log_{10}r + 10.9 - \mathrm{C},$$

where

r = distance from the emitter (m)

$L_{\mathrm{w}} = 10^{-9}$ watts reference level

C = an air density correction term.

The English units algorithm is:

$$L_{\mathrm{W}} = Lp + 20\log_{10}r + 0.6 - \mathrm{C},$$

where

r = distance from the emitter in feet.

C varies from +0.55 at 1100 millibars, –10°C, to –0.73 at 900 millibars, 55°C. Harris provides a C = 0.6 at 1000 millibars, 20°C.

These algorithms apply to situations in which sound can radiate in a free (unobstructed) field. An enclosure that surrounds a machine is anything but a free field. Therefore, sound reverberates inside an enclosure and "builds up."

Estimating reverberative buildup with calculations is arduous and error prone. Instead of attempting to calculate reverberative buildup, we develop a design by degrading the calculated transmission losses for frequencies <1kHz by 6 dB each. Then, we construct a mock-up enclosure and change the thickness or type of liner material as measurements on the mock-up indicate.

9.4 NOISE ENCLOSURE DESIGN RECOMMENDATIONS

9.4.1 In General

1. Use a calibrated sound meter to obtain overall dB(A) sound levels and overall full or fractional octave data. Then identify the high level emitters and gather the same data at those loci.
2. Rank significant emitter levels from greatest to least dB levels.
3. Focus effort on emitters that are within 10 dB of the overall sound level, and do whatever is practical to quiet each.
4. Select a surface mass density that causes the transmission losses (each minus 6 dB) <0.75 kHz to meet design goals.
5. Initially line the interior enclosure surface with thick absorptive material.
6. Assure that the total area of all unsealed openings through the enclosure does not exceed 1% of the total enclosure surface area.
7. Iterate the design to optimize tradeoffs of weight, noise reduction, and component accessibility.

9.4.2 Recurrent Design Elements

Over several enclosure design efforts, (1) sufficient low frequency transmission loss was usually obtained with 1–2 lb/ft2 surface material; and (2) an absorptive liner consisting of 1–3 inches of glass fiber usually was needed.

On unquieted engine-powered items, noise emitted from the engine fan, crankcase, engine air intake, hydraulic components, or the power train was usually within 10 dB of the overall noise level. Exhaust noise was usually not within 10 dB.

9.4.3 Tips

- For early design mock-ups, a low cost enclosure can be made of common 0.5-inch thick gypsum wallboard mounted and sealed with duct tape. To reduce reverberative buildup, the inside should be lined with glass fiber.

- Commonly available spreadsheet software can be used to aid calculation.

REFERENCES

American National Standards Institute. (1983). *Specification for sound level meters (ANSI S1.4-1983).* Standards Secretariat, Acoustical Society of America.

Fletcher, H., & Munson, W. A. (1933). Loudness, its definition, measurement, and calculation. *Journal of the Acoustical Society of America*, 5, 82–108.

Harris, C. M. (Ed.). (1991). *Handbook of acoustical measurements and noise control* (3rd ed.). McGraw-Hill.

Hirshorn, M. (1989). *Noise control reference handbook*. Industrial Acoustics Co.

Kryter, K. D. (1970). *The effects of noise on man*. Academic Press.

Lord, H. W., Gatley, W. S., & Evensen, H. A. (1980). *Noise control for engineers*. McGraw-Hill.

Price, G. R., Kalb, J. T., & Garinther, G. R. (1989). Toward a measure of auditory handicap in the Army. *Annals of Otology, Rhinology & Laryngology*, 98(Suppl.), 140.

FURTHER READING

Balážiková, M., Pačaiová, H., & Tomašková, M. (2023). A proposal for risk assessment of low-frequency noise in the human–machine–environment system. *Applied Sciences*, 13(24), 13321.

Crossley, E., Biggs, T., Brown, P., & Singh, T. (2021). The accuracy of iPhone applications to monitor environmental noise levels. *The Laryngoscope*, 131(1), E59–E62.

King, E. A. (2022). Here, there, and everywhere: How the SDGs must include noise pollution in their development challenges. *Environment: Science and Policy for Sustainable Development*, 64(3), 17–32.

Otterbring, T., Bodin Danielsson, C., & Pareigis, J. (2021). Office types and workers' cognitive vs affective evaluations from a noise perspective. *Journal of Managerial Psychology*, 36(4), 415–431.

Pretzsch, A., Seidler, A., & Hegewald, J. (2021). Health effects of occupational noise. *Current Pollution Reports*, 7, 344–358.

Occupational Noise

10

Measurement and Equipment

E. Tytyk and B. Mrugalska

10.1 INTRODUCTION

According to the Centers for Disease Control and Prevention (CDC) it is estimated that 22 million workers in the United States are exposed to potentially damaging noise at work each year (CDC, 2024). Occupational noise is a significant environmental health hazard, with wide-ranging effects on cardiovascular, mental, and general health. Effective monitoring, technological innovations, public education, and robust policy measures are essential to mitigate its impact and improve work conditions. Frequent and comprehensive noise assessments, considering both short-term and long-term effects, are necessary to fully understand and address the health implications of occupational noise (Diaz et al., 2019). Following OSHA's Noise Standard, the employer must reduce noise exposure through engineering controls, administrative

DOI: 10.1201/9781003675761-10

controls, or Hearing Protection Devices (HPDs). For this aim, hearing conservation programs are developed to prevent occupational hearing loss, preserve and protect remaining hearing, and provide workers with the information and tools they need to protect themselves. Employers must measure noise levels, offer free yearly training, hearing protection, and exams, and assess how well the current hearing protectors work (unless adjustments to tools, equipment, and schedules cause worker noise exposure levels to fall below the 85 dBA threshold) (OSHA, 2025).

10.2 NOISE MEASUREMENT METHODS

Noise measurements are performed to determine the values characterizing the noise emission of machines, which are considered separate sources of noise in experimental or operating conditions, or to assess noise exposure at workstations and in specific places.

The noise emission of the machine can be an important indicator of its technical condition, similar to mechanical vibration parameters. Thus, following European directives, the sound power level of the source or the sound pressure level of the emission should be analyzed. Sound power (W) is the total sound energy radiated by a sound source and measured in watts (W), while sound pressure (P) is the change in the local air pressure when a sound wave passes through. It is important to remember that the sound energy radiating from the source will raise the sound pressure level in the room. The sound power level of the source is independent of the room, but the sound pressure levels will depend on our distance from the source and the characteristics of the room. In order to assess the sound power, we can use the following equation:

$$L_w = 10 log_{W_0}^{W} dB\left(W_0 = 1\,pW\right) \tag{1}$$

where W is the sound power, and W_0 is the reference sound power of 1 picowatt.

The sound pressure level can be obtained:

$$L_p = 10 log_{p_o^2}^{p_0^2} dB\left(p_0 = 20\,\mu Pa\right) \tag{2}$$

where p is the sound pressure, and p_o is the reference sound pressure of 0.00002 pascals = 0 dB the threshold of hearing, in air at 1 kHz.

Sound power can be assessed through either sound pressure or sound intensity. To derive sound power from measurements of sound pressure, two techniques are available: the direct approach and the comparison approach. The direct method is primarily applied in free-field settings, though it can also be employed in reverberant environments. On the other hand, the comparison method is exclusively applicable to reverberant sound fields. In the direct method, an imaginary boundary encloses the device being tested, and sound pressure is measured at multiple locations along this boundary. These measurements are then averaged spatially and adjusted for any influence of the acoustic surroundings, including background noise. After determining the average sound pressure level on the measurement surface, the sound power level can be obtained by scaling for the surface area compared to a reference area of 1 m^2. This is where the sound power level is equal to the sound pressure level.

Knowing the sound power level of a device is very useful, as it allows us to objectively compare the sound output of different devices, without the environment in which it was tested or the distance at which measurements were taken. Therefore, the sound power levels are ideal for product labeling, for specifying noise emission limits for devices, and for verifying compliance with limits. Thus, the sound power level is given in the technical documentation of the device when the average sound pressure level of the emission, corrected by the frequency response A, exceeds 85 dBA.

Measuring noise levels and workers' noise exposures is the most important part of a workplace hearing conservation and noise control program. The results of noise measurements are primarily used to compare existing acoustic conditions with the conditions specified by hygiene standards and regulations (NDN), as well as to assess and select noise reduction measures. During such measurements, the microphone should be placed in the location where the employee's head is usually located. Measurements should be taken in the absence of the employee, as their body may disturb the acoustic field. The microphone should be located at a distance of no less than one meter from a wall or other surface that strongly reflects sound waves (e.g. from windows – at a distance of 1.5 m) and at a height of 1.2 m above the floor (head position in a sitting position) or more (for a standing position, depending on height). The following methods of noise measurement are described in standards PN-N-01307:1994, ISO 1999:2013, and EN ISO 9612:2011:

- direct,
- indirect,
- sampling,
- statistical distribution.

Measurements of quantities characterizing noise may be taken during the working day in selected periods of typical exposure (sampling method) or during the performance of specific tasks and activities. The mode and frequency of measurements, the method of recording and storing results, and the method of making them available to employees are specified in the relevant ministerial regulation. In order to assess the degree of human exposure to noise, the following acoustic parameters are measured (PN-N-01307:1994):

- sound level A – sound pressure level corrected according to the frequency response A – L_A [dB_A],
- sound level C – sound pressure level corrected according to the frequency response A – L_C [dB_C],
- maximum sound level A – the maximum effective value of the sound level A, occurring during the observation – L_{Amax} [dB_A],
- peak sound level C – the maximum instantaneous value of the sound level C, occurring during the observation – $L_{C\,peak}$ [dB_C],
- noise exposure – this is the acoustic energy received by a person in a specific time (usually 8 h) – $E_{A,T}$ [$Pa^2 \cdot s$],
- equivalent sound level – $L_{Aeq,T}$ [dB_A].

Depending on the time characteristics of the noise, an appropriate method of measuring the quantities characterizing the noise should be used. Measurements can be performed during the working day in selected periods of typical exposure (using the sampling method) or during specific tasks and activities. In the case of continuous noise, i.e. when the difference between the maximum and minimum sound pressure level measured in different time intervals is not greater than 5 dB_A, it is sufficient to perform several measurements (at least three) using the time function S (slow) and calculate the arithmetic mean of these results. This is the so-called indirect method of performing measurements. It is characterized by simplicity and speed, but requires certain – although simple – calculations. In such an approach it should be noticed that logarithmic numbers are averaged. From a mathematical point of view, this is an incorrect operation, but the error made in calculating the actual values characterizing noise is small and permissible. The average value calculated in this way can be assumed as an equivalent value of the sound pressure level L_{eq} and compared with the permissible value appropriate for a specific type of work (e.g. work with high intellectual load, sheltered work station) or environment, as included in PN-87/B-02151/02.

Variable noise occurs when there is a specified number of clearly distinguishable sound levels with a difference of at least 5 dB_A. In such a case, the arithmetic mean of several momentary measured values of noise levels cannot be used, because the calculation error in relation to the actual value would be unacceptably large. In such a case, the direct method should be used. It consists in continuous measurement of the employee's exposure to noise using the time function S (slow) and reading the values directly from meters, e.g. a dosimeter or an integrating sound level meter. This is a simple method that does not require complicated calculations, as it uses the function of integrating sound intensity values, built into the measuring system. It minimizes the risk of making measurement errors in the case of variable noise. The disadvantage of this method is that it is time-consuming, as conducting a thorough test takes at least an entire work shift. The value read from the meter is the equivalent sound intensity level corrected with the frequency filter A, determined for 8 hours of noise exposure ($L_{eq,8h}$) and is the basis for assessment by comparison with the permissible value (NDN). Road, rail, and aircraft noise can be measured in a similar way, but the statistical distribution method is also used. It allows for grouping the measured sound levels into classes that differ by 5 dB, 2.5 dB, or 1 dB. Impulsive noise consists of a series of acoustic events lasting a maximum of one second. The duration of a noise impulse is determined by the so-called reverberation time, resulting from the acoustic absorption of materials in the room (this is the time after which the sound level in the room drops by 60 dB). Measurements are performed using the "Impulse" time constant, corresponding to the ear's reaction times to impulse and impact sounds, which are 35 ms for the increase and 1.5 s for the decrease. Currently, the level values for this time constant are not standardized. The Regulation of the Minister of Economy and Labour of August 5, 2005 (item 1318) on occupational health and safety in work involving exposure to noise or mechanical vibrations introduces, in accordance with Directive 2002/44/EC, an additional criterion for assessing daily noise exposure – the Ad(8) action threshold. The noise action threshold values are defined as follows:

- $L_{ex,8h} < 85$ dBA,
- $L_{Amax} < 115$ dBA,
- $L_{Cpeak} < 135$ dB.

The standard also recommends maximum noise levels at the workplace to ensure uninterrupted work performance:

- L_{Aeq} (equivalent sound level) ≤ 55 dBA – in rooms: administrative offices, design offices, for theoretical work, data processing, and others with similar purposes.
- L_{Aeq} (equivalent sound level) ≤ 65 dBA – in the dispatcher, observation, and remote control cabins with telephone communication used in the control process, in rooms for performing precision work, and in other rooms with similar purposes.

- L_{Aeq} (equivalent sound level) ≤ 75 dBA – in direct control cabins without telephone communication, in laboratories with noise sources, in rooms with counting machines and devices, typewriters, teleprinters, and in other rooms with similar purposes.

If these values are above, the employer must take action to reduce the occupational risk and implement a hearing conservation program.

Performance of work in continuous noise levels of 95–100 dB cannot exceed 40–100 minutes per day, while work in noise levels up to 110 dB cannot exceed 10 minutes per day. In Poland, the permissible noise levels for minors and pregnant women are specified in the Regulation of the Council of Ministers of August 24, 2004, on the list of work prohibited for minors and the conditions of their employment in some of these works, and in the Regulation of the Council of Ministers of April 3, 2017, on the list of arduous, hazardous, or harmful work for the health of pregnant women and breastfeeding mothers.

If, during the last two noise measurements conducted within a two-year interval, the intensity of the factor did not exceed 0.2 of the NDN value, the employer may refrain from conducting further measurements. However, the above tests and measurements are repeated when there have been changes in technical equipment, technological processes, or working conditions that could have affected the level of emissions, and exposure levels, or when circumstances arise that justify their repetition (Kołodziejczyk, Leśniak, 2021).

10.3 MEASUREMENT EQUIPMENT

The most common instruments used for measuring noise are the sound level meter (SLM), the integrating sound level meter (ISLM), and the noise dosimeter. However, in practice, due to diverse environments where noise can appear, it is possible to also find other solutions, as mentioned in Table 10.1.

For measurements of quantities characterizing all types of noise, integrating sound level meters (SLM) and dosimeters with accuracy class 2 or better should be used. During measurements, the dynamic characteristic S (slow) should be included, averaging the results in 0.5 s, and appropriate correction filters: with characteristic A when measuring equivalent levels in order to assess exposure to continuous noise or with characteristic C when measuring peak values of noise. Measuring instruments should be calibrated and standardized before and after measurements. The sensor that responds to quantities characterizing sound is a microphone – for measurement purposes, it is a condenser microphone. It is a technical analogue of the human hearing aid,

TABLE 10.1 Examples of Noise Measurement Equipment

METHOD/EQUIPMENT	*APPLICATION*	*ADVANTAGES*
Acoustic camera	Locating noise sources	Effective tracking
Digital signal processing	EMI noise suppression	Allows measurements in noisy environments
Hydrophone	Underwater noise measurement	Practical for complex environments
Microtremor measurement	Seismic noise recording	Cost-effective
Raspberry Pi system	Seismic noise recording	Low-cost alternative
Real-time digital noise cancellation system	EMI measurements	Removes background noise from audio in real-time while preserving human speech

which is located in the ear. It allows for measuring sound pressure levels in the first accuracy class (according to the IEC 61672:2002) and allows for:

- conducting measurements of noise hazard at workstations and noise in the outdoor environment (in such cases, an important element is the microphone windscreen, e.g. a black sponge ball),
- simplified "Start/Stop" operating mode,
- performing measurements in three parallel profiles,
- sound analysis in octave and third octave bands (to choose from),
- saving the time history of the measurement – Data Logger function,
- saving results in memory on a microSD card (standard 4 GB),
- creating noise maps,
- recording an audio signal (option),
- saving an audio note for the measurement,
- acting as an acoustic dosimeter (option),
- monitoring the device's natural vibrations.

REFERENCES

CDC (2024). About occupational hearing loss. Retrieved on 20 January 2025 from: https://www.cdc.gov/niosh/noise/about/index.html

Diaz J., Asensio C., González S. & Linares C. (2019). A nuanced approach to the environmental noise guidelines for the European Region related to traffic noise. *Environmental Research*, 177, 108582.

ISO 1999:2013, Estimation of noise-induced hearing loss, ISO.

ISO 9612:2009 Acoustics - Determination of occupational noise exposure - Engineering method, ISO.

IEC 61672-1:2002 Electroacoustics - Sound level meters - Part 1: Specifications.

Kołodziejczyk E. & Leśniak G.J. (2021). Hałas w środowisku pracy musi być pod kontrolą. Retrieved on 20 January 2025 from https://www.prawo.pl/kadry/halas-w-srodowisku-pracy,186770.html

PN-N-01307:1994 Hałas - Dopuszczalne wartości hałasu w środowisku pracy - Wymagania dotyczące wykonywania pomiarów. PKN: Warszawa (in Polish).

OSHA (2025). Occupational noise exposure. Retrieved on 20 January 2025 from: https://www.osha.gov/noise

FURTHER READING

Golmohammadi, R., Darvishi, E., Motlagh, M. S., & Faradmal, J. (2021). Role of individual and personality traits in occupational noise-induced psychological effects. *Applied Acoustics*, 173, 107699.

Pretzsch, A., Seidler, A., & Hegewald, J. (2021). Health effects of occupational noise. *Current Pollution Reports*, 7, 344–358.

Roberts, B., Smith, S., Vahora, M., & Miller, E. (2024). Self-reported occupational noise exposure and hearing protection device use among NHANES participants and the risk of hearing loss. *Journal of Occupational and Environmental Hygiene*, 21(9), 623–628.

Shkembi, A. (2024). Caution with self reported occupational noise exposures. *Journal of Occupational and Environmental Hygiene*, 22(2), D3.

Zhang, M., Gao, X., Murphy, W. J., Kardous, C. A., Sun, X., Hu, W., & Qiu, W. (2022). Estimation of occupational noise–induced hearing loss using kurtosis-adjusted noise exposure levels. *Ear and Hearing*, 43(6), 1881–1892.

11 Human Exposure to Vibration

D. E. Wasserman

11.1 INTRODUCTION

Worldwide there are millions of people who go to work each day who are exposed to mechanically induced vibration. These include truck, bus, heavy equipment, fork lift, and farm vehicle operators, train engineers and conductors, miners, helicopter, fixed-wing aircraft, and ship personnel, and operatives at fixed plant sites. These people are exposed to head-to-toe whole-body vibration (WBV). Users of hand-held power tools such as chain saws, and various electrical and air-driven pneumatic tools such as chippers, grinders, sanders, mining tools, drills, and road rippers are exposed to segmental, or hand–arm, vibration (HAV).

DOI: 10.1201/9781003675761-11

11.2 VIBRATORY MOTION

Mechanical vibration can be oscillatory, impact, or random, and each is characterized as a vector quantity and thus consists of both an intensity or magnitude and a direction. Vibration measured at any one point consists of six vectors: three linear mutually perpendicular vectors and three rotational vectors called pitch, yaw, and roll. Generally in human vibration work, only linear measurements are obtained. Root-mean squared (RMS) acceleration is usually the intensity measurement quantity of choice and is expressed in either ms-1 s-1 or gravitational g's, where 1g 9.81 m s-1 s-1. There are specific vibration frequencies (Hz) where WBV is measured (1–80 Hz) and where HAV is measured (5–1500 Hz). Finally, there is a undesirable phenomenon known as resonance or natural frequency where humans are optimally tuned to impinging vibration, resulting in involuntary amplification and exacerbation of its effects at these resonant frequencies; WBV resonance occurs at 4–8 Hz for vertical and 1–2 Hz for the side-to-side or front-to-back directions. HAV resonance is thought to occur at 100–250 Hz (Wasserman 1987).

11.3 HUMAN EFFECTS OF VIBRATION EXPOSURE

The major human effects can generally be subdivided by either WBV or HAV exposure.

11.3.1 Major WBV Effects

At WBV frequencies 1 Hz, sea sickness or kinetosis can occur, which usually ceases when exposure stops. At frequencies 1 Hz, both acute and chronic WBV effects can occur. Generally, if acute exposure occurs at the 4–8 Hz resonance band, there can be a potential safety issue, where, for example, a vehicle operator might not be able safely to control the steering wheel; there also can be discomfort due to WBV exposure (Griffin 1990).

The major medical effects of chronic WBV exposure are principally, but not limited to: (1) severe lower back pain; (2) lumbar spine disc: degeneration, loss of moisture, buckling, slipping/prolapse, herniation, tearing, distortion, excessive pressure (Dupuis and Zerlett 1986; Wilder et al. 1994); and (3) spontaneous abortions and related gynecological problems have been reported in female vehicle drivers chronically exposed to WBV (Abrams 1990).

11.3.2 Major HAV Effects

Chronic HAV exposure has been causally linked for 80 years to a generally irreversible condition of the fingers and hands known as Hand–Arm Vibration Syndrome (HAVS); previously called Vibration White Fingers, and Raynaud's Phenomenon of Occupational Origin. HAVS (Pelmear and Wasserman 1998) is initially characterized by attacks of paresthesia (tingling and/or numbness in the fingers and hands). As HAV exposure continues, finger blanching attacks in one or more fingers, lasting 5–15 minutes, occur particularly in the presence of cold. With continued HAV exposure, this condition is cumulative and progressive, involving multiple fingers; in its terminal stage (rarely seen), gangrene of the fingers can occur, requiring digit amputation. Medical treatment involves the use of so-called "calcium channel blocker" medications, which are not curative but are palliative (i.e. easing pain and suffering without curing). Thus, both early detection of HAVS and the cessation of HAV exposure are essential for these hand power tool operators.

Typically, HAVS prevalences as high as 50% have been found in pneumatic vibrating hand-tool workers in the United States. Finally, it is important to note that a physician's early and careful differential diagnosis is essential to distinguish HAVS from other similar conditions such as idiopathic Primary Raynaud's Disease and (non-vibration) Carpal Tunnel Syndrome.

11.4 VIBRATION HEALTH AND SAFETY STANDARDS

Numerous nations worldwide have participated in the development of consensus health and safety standards/guides in an effort to protect workers against the hazards of both WBV and HAV exposure. Initial standards efforts were begun by the International Standards Organization (ISO) in the late 1960s and continue to this day in cooperation with national standards-setting organizations such as the American National Standards Institute in New York, British Standards Institute in London, and others including the European Union.

Examples of WBV standards/guides include ISO 2631, ANSI S3.18, and BSI 6841. Examples of HAV standards/guides include ISO 5349, ANSI S3.34, and BSI 6842. Worldwide, virtually all WBV standards employ the same measurement coordinate system; the same is true for HAV standards. This permits easy comparison of vibration data made by various groups.

11.5 CONTROLLING VIBRATION EXPOSURE

The major engineering methods of vibration reduction include isolation (the intentional alteration of the pathway between the vibration source and the human receiver) and damping (vibration impinging upon a viscoelastic material causes the material to deform, thus converting mechanical energy into a small amount of heat).

11.5.1 Reducing WBV Exposure

Air-ride seats are commonly used in trucks, buses, farm and construction equipment, ships, fixed plants, etc. to isolate the operator's upper torso/spine from WBV, especially in the 4–8 Hz region. Also in use are isolated truck cabs and special vehicle suspensions. In fixed plant locations, isolating machinery and the remote placing of controls all help to reduce WBV exposure. Administrative controls include regular rest breaks and not lifting cargo immediately after emerging from driving a vehicle.

11.5.2 Reducing HAV Exposure

First, workers are advised to use tools that are both anti-vibration (A/V) designed and ergonomically designed, since an ergonomically designed tool may only represent the optimum interface between the worker and the tool (i.e. handle design, weight, etc.), but these tools do not necessarily have reduced vibration characteristics within the tool *per se*. Second, do not, as a rule, use "tool wraps" that use a wade of vibration damping material merely wrapped around a conventional tool handle, because of both the increased handle diameter and the limited ability of the wrap to damp HAV from certain tools.

Third, use only full-finger, good fitting, A/V gloves that reduce vibration and keep the fingers and hands warm and dry. Do not use "exposed fingers" gloves that only protect the palm, since HAVS nearly always begins at the fingertips. Fourth, good work practices include: keeping the entire body warm and dry especially the hands; no smoking while using vibrating hand-tools, since nicotine, cold, and vibration all constrict blood flow. Let the tool do the work, grasping it as lightly as possible, consistent with safe work practices. If signs and symptoms of HAVS appear, see a physician immediately. Keep the tool and implements in good working order; replace worn-out tools promptly.

REFERENCES

Abrams, R. (1990). Sound and vibration in pregnancy. In R. Creasy & J. Warshaw (Eds.), *Seminars in perinatology, part II* (pp. 273–334). W.B. Saunders.

Dupuis, H., & Zerlett, G. (1986). *The effects of whole-body vibration*. Springer.

Griffin, M. (1990). *Handbook of human vibration*. Academic Press.

Pelmear, P., & Wasserman, D. (1998). *Hand–arm vibration: A comprehensive guide for occupational health professionals* (2nd ed.). OEM Press.

Wasserman, D. E. (1987). *Human aspects of occupational vibration*. Elsevier.

Wilder, D., Wasserman, D., Pope, M., Pelmear, P., & Taylor, W. (1994). Vibration. In P. Wald & G. Stave (Eds.), *Physical and biological hazards of the workplace* (pp. 64–83). Van Nostrand Reinhold.

FURTHER READING

Mocan, A., Gaureanu, A., Szabó, G., Mrugalska, B. (2022). Arguments for emerging technologies applications to improve manufacturing warehouse ergonomics. In A. Draghici & L. Ivascu (Eds.). *Sustainability and innovation in manufacturing enterprises. Advances in sustainability science and technology*. Springer. https://doi.org/10.1007/978-981-16-7365-8_5

Oroszi, T., Geerts, E., de Boer, S. F., Schoemaker, R. G., van der Zee, E. A., & Nyakas, C. (2022). Whole body vibration improves spatial memory, anxiety-like behavior, and motor performance in aged male and female rats. *Frontiers in Aging Neuroscience*, 13, 801828.

Trama, R., Hautier, C., Blache, Y., Bertucci, W., Chiementin, X., & Hintzy, F. (2023). Intra-cycle analysis of muscle vibration during cycling. *Sports Biomechanics*, 22(4), 554–566.

Welch, D., Shepherd, D., Dirks, K. N., & Reddy, R. (2023). Health effects of transport noise. *Transport Reviews*, 43(6), 1190–1210.

Zhou, X., Liu, Y., Luo, M., Zheng, S., Yang, R., & Zhang, X. (2022). Overall and thermal comfort under different temperature, noise, and vibration exposures. *Indoor Air*, 32(1), e12915.

Index

For Product Safety Concerns and Information please contact our EU representative GPSR@taylorandfrancis.com Taylor & Francis Verlag GmbH, Kaufingerstraße 24, 80331 München, Germany

Printed and bound by CPI Group (UK) Ltd, Croydon, CR0 4YY
16/12/2025
02020517-0011